PARASITIC PROTOZOA

Biological Sciences

Editor
PROFESSOR A. J. CAIN
MA, D.PHIL
Professor of Zoology
in the University of Liverpool

PARASITIC PROTOZOA

J. R. Baker, D.Sc.
Senior Lecturer in the Department of Medical Protozoology,
London School of Hygiene and Tropical Medicine,
University of London

HUTCHINSON UNIVERSITY LIBRARY
LONDON

HUTCHINSON & CO (*Publishers*) LTD
178–202 Great Portland Street, London W1

London Melbourne Sydney
Auckland Bombay Toronto
Johannesburg New York

First published 1969

The paperback edition of this book is sold subject to the condition that it shall not, by way of trade or otherwise, be lent, re-sold, hired out, or otherwise circulated without the publisher's prior consent, in any form of binding or cover other than that in which it is published and without a similar condition including this condition being imposed on the subsequent purchaser

© J. R. Baker 1969

This book has been set in Times, printed in Great Britain on Smooth Wove paper by Anchor Press, and bound by Wm. Brendon, both of Tiptree, Essex

09 099161 3
09 099160 5

Dedicated to
Professor P. C. C. Garnham, C.M.G., F.R.S.
Emeritus Professor of Medical Protozoology
in the University of London

CONTENTS

	Preface	9
	Introduction	11
1	Classification and evolution of the parasitic Protozoa	15
2	Anatomy and physiology of the Protozoa	27
3	Trypanosomes and related parasites	40
4	Parasitic flagellates of the alimentary and urinogenital tracts	64
5	Parasitic amoebae	76
6	Gregarines, Coccidia and Haplosporea	87
7	Malaria parasites and their relatives	100
8	Piroplasms	117
9	Toxoplasmea	127
	Addendum: *Pneumocystis carinii*	135
10	Cnidospora	137
11	Parasitic ciliates	145
12	Techniques for intestinal parasites	152
13	Techniques for tissue parasites	158
	References	165
	Index	171

PREFACE

This book attempts to give an introduction to a systematic study of the parasitic Protozoa. Although it inevitably reflects its author's interests in being biased towards the organisms of medical and veterinary importance, it is hoped that the other parasitic Protozoa have been dealt with in sufficient detail to present a coherent picture of the group as a whole. Ideally, the parasitic members of the phylum should be considered in the context of their more numerous non-parasitic relatives; this has not been attempted, but readers are referred to books where it has been done—notably the classic work by Wenyon (1926), now sadly outdated but never surpassed. More recently, the Protozoa as a whole have been dealt with by Mackinnon and Hawes (1961), Manwell (1961), Dogiel (1965), Kudo (1966) and, except for the ciliates, Grassé (1952, 1953), and these books may be used to supplement the account given below.

The general outline of this book is based on the course of lectures given in the Department of Medical Protozoology of the London School of Hygiene and Tropical Medicine, and I obviously owe much to my colleagues in that department for the many critical discussions which have taken place—particularly during the writing of this book. Special credit must be given to Professor P. C. C. Garnham, director of the department until September 1968 and my kindly initiator into, and guide through, the fascinating complexities of parasitological Protozoology (or protozoological Parasitology). Mr R. Killick-Kendrick contributed greatly, especially to the material contained in Tables 2 and 3. Dr R. S. Bray deserves my thanks for his stamina in reading the entire typescript and making corrections and constructive criticisms. The responsibility for the inevitable remaining errors is, of course, mine. I also wish to thank Mrs R. E. Eversden for preparing the typescript and Miss J. Freeman for the index.

The illustrations have all been drawn especially for this book (except for Fig. 127), mostly from slides in my collection or that of the Department of Medical Protozoology. All those which illustrate organisms (except Figs. 67, 72, 73 and 131) were made from stained specimens.

Finally, although I am clearly in no position to deny the truth of the proverb 'Little things please little minds', I should like to declare my affection for the organisms described in this book, which is not entirely due, I hope, to the fact that they provide me with my livelihood. I hope some readers will be able to share this affection.

INTRODUCTION

In the first chapter of this book an attempt is made to answer the question 'What is a protozoon?'. In this introduction it is necessary to consider briefly the related question 'What is a parasite?', and to define some terms.

The phenomenon of parasitism has been discussed at length by Sprent (1963) and Dogiel (1964); here it will be considered only very briefly. The *Concise Oxford English Dictionary* (fourth edition, 1951) defines a parasite as an 'animal ... living in or upon another and drawing nutriment directly from it', and derives the word 'parasite' from the Greek *para* ('beside, beyond, wrong, irregular') and *sitos* ('food'). In zoology, *parasitism* may be broadly defined as an association between two animals of such a kind that one lives and feeds,[1] temporarily or permanently, either in or on the body of the other. It will be noted that this definition includes the mammalian foetus, and it is probably fair to regard this association as an example of a rare phenomenon, *intra-specific parasitism* (also exemplified by the ectoparasitic males of certain Crustacea (barnacles) and deep-sea angler fish). Parasitism between animals belonging to different species can be referred to as *inter-specific parasitism*.

In any parasitic association the smaller of the associating pair of animals is regarded as the *parasite* and the larger as the *host*. According to whether the parasite lives in or on its host's body, the association is referred to as *endo-* or *ectoparasitism* respectively. Some

---

[1] An association between two animals in which one is merely transported from place to place on the body of the other, without feeding while thereon, is called *phoresy* (from the Greek *phero*, to bear).

parasites, or some stages in the life cycle of a parasite, are unable to live apart from their hosts: they are called *obligate* parasites. Others can live equally well free: these are *facultative* parasites.

Parasitism can be broadly divided into two types of association on the basis of whether only one partner (the parasite) benefits, or whether both parasite and host do so. The former situation is subdivided by Sprent (1963) into true parasitism and commensalism. *Commensalism* (Latin *com-*, together, and *mensa*, a table) is defined as an association in which the parasite does *not* feed on the host's tissues; such a parasite is known as a commensal organism. Examples of both ecto- and endocommensalism are known. Certain ciliates live as ectocommensals on the body surface of fish. The majority of the intestinal parasites of man and other animals are endocommensals. Their hosts are only marginally affected, if at all, by the fact that the parasite diverts some of the host's food to its own use. True parasitism is, by contrast, an association in which the parasite feeds on the tissues of the host; unfortunately there is no single word for it. Some people restrict the use of the word parasitism to this particular sense, while others use it, as it is being used in this book, more broadly. Parasitism in Sprent's sense may perhaps be differentiated as *tissue parasitism*. Not all tissue parasites are necessarily, or always, harmful to their hosts: many trypanosomes (e.g. *Trypanosoma lewisi*) are undoubtedly tissue parasites (they feed on blood plasma) but apparently do no harm to their hosts; others may be exceedingly harmful, 'even unto death' (e.g. *T. brucei rhodesiense* in man). Such harmful parasites are said to be *pathogenic*. Some parasites (in the broad sense) live sometimes as commensals and at other times as tissue parasites (e.g. *Entamoeba histolytica* in man).

The other broad division of parasitism, in which both partners benefit, is called *symbiosis* and the partners *symbionts* (Greek *syn-* or *sym-*, together, and *bios*, life). Proven examples are rare among Protozoa, but a classic one is the association between hypermastigid flagellates and termites, in which the flagellates, unable to survive outside the gut of their host, digest the cellulose on which the termite exclusively feeds and which it itself is entirely unable to digest; thus without its flagellates, the termite dies of starvation no matter how abundant its food supply (see Chapter 4). The relationship bebetween the fungal and algal constituents of lichen is another classic example of symbiosis. Protozoan symbionts are, as far as is known, always endoparasitic, but symbiotic relationships involving ectoparasites are known amongst other phyla (e.g. sea anemones living on the shells of hermit crabs, the former benefiting from the provision of transport and particles of food dropped by the crab and the latter being defended against predators by the cnidoblasts of their passengers).

Many parasites have only a single host during their life cycle, part of the latter being spent outside the host; others have two, occasionally more, hosts, usually belonging to widely separated taxonomic groups. The two hosts are sometimes distinguished as the *definitive host* (in which the parasite undergoes sexual reproduction) and the *intermediate host* (in which it does not). This distinction is, however, impossible with parasites like the trypanosomes, which have no sexual process, and inappropriate with those like the malaria parasites, in which the 'intermediate' host (the vertebrate) is probably the one in which the parasites' ancestors evolved, the second 'definitive' host (an insect) being adopted only relatively late in the evolutionary process (see Chapter 1). Thus, at least in protozoology, these terms are seldom used. One of the hosts is, instead, often referred to as the *vector,* a term which is difficult to define objectively but which implies that the host so named transmits the parasite to the other host. From the human point of view the mosquito is the vector of malaria; but from the mosquito's viewpoint, man is the vector. In practice, the term vector is restricted to the invertebrate host when the other host is a vertebrate (or, rarely, a plant). If the parasite undergoes part of its life cycle in the vector, its transmission is said to be *cyclical;* if not, it is referred to as *non-cyclical* or 'mechanical' transmission.

I

CLASSIFICATION AND EVOLUTION OF THE PARASITIC PROTOZOA

CLASSIFICATION

The Protozoa are regarded as a phylum within the animal kingdom. They include the smallest animals (though the largest Protozoa are bigger than the smallest Metazoa) and are often said to be the most primitive. This is true inasmuch as they probably differ the least from the original, hypothetical group of living organisms that was ancestral to both the plant and animal kingdoms, but the Protozoa have, of course, been evolving for just as long as we have, and some of them are very highly evolved animals indeed. The fact that the Protozoa are the modern representatives of the group which was ancestral to both animals and plants leads to taxonomic difficulties. Among the Mastigophora (flagellates), some forms possess chloroplasts and have typically plant-like (holophytic) nutrition; others, obviously closely related, lack chlorophyll and have typically animal-like (holozoic) nutrition. Thus constant verbal battles are fought between botanists and zoologists for jurisdiction over these groups. The suggestion made by Haeckel in 1866, that the two groups should not be separated but included, with all the other Protozoa and the unicellular Algae, in a third kingdom, the Protista, is basically sound, but it has not been generally adopted, perhaps because it does not appeal to the 'pigeon-holing' mentality of most taxonomists.[1] The best practical solution seems to be to retain the green flagellates

[1] This is not intended as a criticism of taxonomists: such a mentality is probably a prerequisite for taking up taxonomy, a subject which—though often maligned by the so-called 'experimental' biologists—is basic to all other biological disciplines. But it is, perhaps, important to remember that taxonomists are often attempting the impossible,

in both the Algae and the Protozoa, while remembering that this is only an approximation to the truth.

On this basis the Protozoa may be regarded as a group of animals which have not adopted multicellular somatic organization. This circumlocution avoids the necessity of having to state categorically that the Protozoa are unicellular animals (as most authors do), or that they are non-cellular (as a few authors do). As in many controversies of this sort, both concepts are true; it depends on the angle from which one views the question. Structurally the Protozoa are undoubtedly unicellular organisms: the anatomy of a Protozoon is basically identical with that of a single metazoan cell (see Chapter 2). Functionally, however, a protozoon is undoubtedly non-cellular in the sense that it is a whole, complete animal—just as much as an elephant or a man is—but it has not adopted the expedient of dividing itself into a large number of small structural units or cells. One result of this has been to limit the size of Protozoa: an amoeba the size of an elephant would be rather unwieldy. Another consequence has been to limit the extent to which Protozoa have been able to become independent of their environment and, in this sense, the Protozoa are indeed primitive animals. In terms of the number of individuals alive at any given time, however, they are probably the most successful of all animals.

Not surprisingly (see footnote on p. 15 above), there is no complete agreement as to how the Protozoa should be classified. The scheme proposed by a committee of the Society of Protozoologists (Honigberg *et al.*, 1964) represents the nearest to a consensus which is available, and it is this scheme which is adopted in this book—with one important, and a few minor, exceptions. The piroplasms (Chapter 8), which for many years had been regarded as close relatives of the malaria parasites (subphylum Sporozoa, class Teleosporea, subclass Coccidia, order Eucoccida, suborder Haemosporina), were removed from this subphylum by the committee and placed with the amoebae (subphylum Sarcomastigophora, superclass Sarcodina) as a separate class, Piroplasmea. The evidence for this move was, on the whole, negative (i.e. the absence of sporozoan features rather than the presence of sarcomastigophoran ones), although it is doubtless true that the piroplasms are less closely related to the malaria parasites than had been thought. Recent studies with the electron microscope have revealed more and more clearly an anatomical

---

trying to impose arbitrary divisions on that which is in fact a continuum. Thus the perfect classification is a myth and its pursuit, however valuable, a labour of Sisyphus. In other words, taxonomists will never be out of work.

Classification and evolution of the parasitic Protozoa

relationship between the piroplasms and the Sporozoa, so in this book they will be treated as a separate class within that subphylum (p. 19).

The classification adopted in this book is, therefore, that set out below. The main diagnostic features of each group are given in parentheses and the names of important parasitic genera are included (in italics). The classification of the groups containing parasites is given in more detail.

PHYLUM PROTOZOA

Subphylum I. Sarcomastigophora (locomotion by flagella, pseudopodia,[1] or both)

SUPERCLASS 1. Mastigophora ('flagellates': locomotion mainly or entirely by flagella; division symmetrogenic)

CLASS 1. Phytomastigophorea (with chlorophyll; none parasitic)

CLASS 2. Zoomastigophorea (without chlorophyll)

Order Rhizomastigida (pseudopodia and flagella present simultaneously or at different phases of life cycles; mostly non-parasitic) *Histomonas* (?), *Dientamoeba*

Order Kinetoplastida (possess kinetoplast)

Suborder Bodonina (usually two flagella; mostly parasitic) *Cryptobia*

Suborder Trypanosomatina (one flagellum; all parasitic) *Leptomonas, Phytomonas, Leishmania, Trypanosoma,* etc.

Order Retortamonadida (2–4 flagella; cytostome bordered by fibril; all parasitic) *Retortamonas, Chilomastix,* etc.

Order Diplomonadida (bilaterally symmetrical, with two similar nuclei and four pairs of flagella; mostly parasitic) *Giardia, Hexamita,* etc.

Order Oxymonadida (4 or more flagella, typically in two pairs, and one or more nuclei and axostyles; no cytostomal fibril; no parabasal body; all parasitic in termites or insects)

[1] For definitions and explanations of terms, see Chapter 2.

Order Trichomonadida (4–6 flagella, one trailing and often with undulating membrane; parabasal body; all parasitic) *Trichomonas*

Order Hypermastigida (numerous flagella; all parasitic in termites or insects)

   Suborder Lophomonadina (organelles arranged in single system) *Lophomonas*

   Suborder Trichonymphina (organization basically bilateral) *Trichonympha*

Two other orders (non-parasitic)

SUPERCLASS 2. Opalinata (numerous short flagella; 2 to many similar nuclei; all parasitic, usually in amphibia)

Order Opalinida; *Opalina*

SUPERCLASS 3. Sarcodina ('amoebae'; locomotion mainly or entirely by pseudopodia)

CLASS 1. Rhizopodea (pseudopodia are not axopods)

   SUBCLASS 1. Lobosia (pseudopodia are lobopods)

Order Amoebida (naked, i.e. without 'shell'; many parasitic) *Entamoeba, Endamoeba, Endolimax, Iodamoeba, Hartmanella*

One other order (non-parasitic)

Four other subclasses (non-parasitic)

One other class (non-parasitic)

Subphylum II. Sporozoa (typically produce simple, resistant spores, containing one or more sporozoites; no cilia or flagella, except on male gametes of some; pseudopodia rarely formed; all parasitic)

CLASS 1. Telosporea (all have sexual reproduction; spores present in most, and sporozoites in all)

   SUBCLASS 1. Gregarinia ('gregarines'; mature trophozoites large, extracellular; parasites of gut and body cavity of invertebrates)

Order Eugregarinida (no schizogony) *Monocystis,* etc.

Order Schizogregarinida[1] (schizogony) *Selenidium,* etc.

[1] The Committee's classification recognizes three orders—the Archigregarinida, Eugregarinida and Neogregarinida. This point is discussed on pp. 89–90 below.

Classification and evolution of the parasitic Protozoa

SUBCLASS 2. Coccidia (mature trophozoites small, intracellular)

Order Protococcida (no schizogony; in marine annelids)

Order Eucoccida (schizogony; asexual and sexual phases of multiplication)

Suborder Adeleina (spores in some genera; male and female gametocytes develop in association—syzygy; one or two hosts) *Adelea, Haemogregarina, Hepatozoon, Karyolysus, Klossia, Klossiella*

Suborder Eimeriina (spores in some genera; no syzygy; non-motile zygote; one or two hosts) *Eimeria, Isospora, Lankesterella,* etc.

Suborder Haemosporina (no spores; no syzygy; motile zygote; two hosts, asexual development in a vertebrate, sexual completed in a dipterous insect) *Plasmodium, Haemoproteus, Hepatocystis, Leucocytozoon,* etc.

CLASS 2. Piroplasmea (no spores; small, non-pigmented parasites of erythrocytes and sometimes other cells of vertebrates; also invertebrate host; reproduce by schizogony or binary fission, presence of sexual process uncertain)

Order Piroplasmida; *Babesia, Theileria, Cytauxzoon, Dactylosoma*

CLASS 3. Toxoplasmea (no spores; no sexual reproduction known; cysts or pseudocysts containing many naked zooites; multiply by endodyogeny and, possibly, binary fission)

Order Toxoplasmida; *Toxoplasma, Sarcocystis, Besnoitia*

CLASS 4. Haplosporea (spores; schizogony; no sexual reproduction known; single host)

Order Haplosporida

Subphylum III. Cnidospora (spores with one or more polar filaments; all parasitic; single host)

CLASS 1. Myxosporidea (spores multicellular)

Order Myxosporida (spore with one or two sporoplasms and two—rarely one to six—polar capsules containing coiled polar filament; in cold-blooded vertebrates)

Suborder Bivalvulina¹ (2 shell valves) *Myxobolus,* etc.

Suborder Multivalvulina¹ (more than two shell valves) *Kudoa,* etc.

Order Actinomyxida (spore with 3 polar capsules, each enclosing polar filament; several to many sporoplasms; in invertebrates, mainly annelids)

Order Helicosporida (spore with 3 sporoplasms surrounded by thick, spirally coiled polar filament; in insects)

CLASS 2. Microsporidea (spore unicellular, with one polar filament and one sporoplasm; in invertebrates and vertebrates)

Order Microsporida

Suborder Monocnidina (single, independent spores) *Nosema, Glugea, Thelohania, Plistophora,* etc.

Suborder Dicnidina (spores united in pairs)

Subphylum IV. Ciliophora (cilia present at some or all stages of life cycle; nuclei of 2 dissimilar kinds; binary fission homothetogenic; sexual reproduction usually by conjugation)

CLASS Ciliatea

SUBCLASS 1. Holotrichia (body ciliature simple and uniform in most orders, special buccal cilia absent or inconspicuous)

Order Gymnostomatida (cytostome on body surface; some parasitic in ruminants or equids) *Buetschlia,* etc.

Order Trichostomatida (cytostome at base of vestibule; some parasitic) *Isotricha, Dasytricha, Balantidium*

Order Chonotrichida (no body cilia, mouth cilia in apical 'funnel'; ectoparasitic on Crustacea)

Order Hymenostomatida (cilia of oral region fused to form membranelles: few parasitic) *Tetrahymena, Ichthyophthirius*

Three other orders (non-parasitic)

[1] This classification differs from the Committee's and is based on that of Shulman (1964).

Classification and evolution of the parasitic Protozoa

SUBCLASS 2. Peritrichia (body ciliature essentially absent; conspicuous oral ciliature, winding anti-clockwise; body often attached to substrate by adhesive disc or stalk which may be contractile; some ecto-parasitic on aquatic hosts)

Order Peritrichida; *Trichodina,* etc.

SUBCLASS 3. Suctoria (adults with no cilia, usually attached to substrate by non-contractile stalk; feed by means of sucking tentacles; some ectoparasitic on aquatic hosts, 1 endoparasitic in gut of equids)

Order Suctorida; *Allantosoma*

SUBCLASS 4. Spirotrichia (body ciliature usually sparse; conspicuous oral ciliature, winding clockwise)

Order Heterotrichida (body ciliature usually present and uniform; few endoparasitic forms and some—in separate suborder, lacking body cilia—ectoparasitic on marine hosts) *Nyctotherus*

Order Entodiniomorphida (body cilia reduced to tufts, or absent; firm pellicle, often with posterior spines; parasitic in herbivores) *Entodinium, Diplodinium, Ophryoscolex,* etc.

4 other orders (non-parasitic)

EVOLUTION

Discussion on the evolution of the Protozoa is inevitably hypothetical, since almost all of them (and certainly all of the parasitic forms) have no bony skeletons or other hard parts which might provide a fossil record of their evolution. Hypotheses of the course of evolution which they have undergone, therefore, have to be based entirely on deduction from the similarities and apparent relationships of the surviving groups only, with the 'missing links' being filled in by (one hopes) judicious guesswork.

It is generally accepted, because of the obvious similarities between the simpler members of both groups (see p. 15 above), that both animal and vegetable kingdoms had their origins in a single group of organisms; and it is widely, if not generally believed that the present-day organisms which have deviated least from this primordial group are the Sarcomastigophora. This belief is based on the facts that (a) flagella (or cilia) are almost universally present amongst plants and animals, and (b) amongst the Mastigophora, species with chloroplasts and species without them are inextricably mixed (to the despair of taxonomists). From this primordial assemblage, evolution presumably proceeded along two main lines: the 'virtuous' organisms which retained chlorophyll and thus synthesized their own food, and the 'lazy' (animal) organisms which adopted the habit of eating the fruits of others' labours (or the others themselves). Thus in a sense all animals are parasitic (but not, of course, within the accepted definition of this term; see p. 11).

Within the animal group, evolution again continued along two main lines (or possibly more, some having become extinct), the flagellate line and the amoeboid line. The present-day survivors of these two lines are the Zoomastigophorea, Opalinata and Ciliophora on the one hand, and the Sarcodina on the other. From these two lines, others branched off at various times. The Sporozoa probably arose from somewhere close to the original flagellate-amoeboid dichotomy, since they possess characters of both groups as well as specializations of their own. The origin of the Cnidospora is more obscure: although conventionally placed near to the Sporozoa they are in fact probably not particularly closely related to this group and are more likely to have arisen separately from the early sarcodine stock (possibly fairly close to the offshoot leading to the metazoan Coelenterata). Ideas about evolution within the various protozoan groups become even more speculative, and will be considered only broadly here. Further discussion (and references) can be obtained from the review by Baker (1965). Within the Ciliophora and Sarcodina, the parasitic habit was probably adopted independently by organisms of several different groups. Almost all endoparasitic members of this group are mainly or exclusively intestinal parasites,[1] and it is easy to imagine how they could have entered the host's alimentary canal in food or water and multiplied there; eventually, mutation and selection would have given rise to organisms better suited to this relatively sheltered existence and, finally, unable to survive outside it unless protected by a cyst. Subsequent evolutionary 'experimentation' would have led to tissue invasiveness, as shown today by what are probably the most highly evolved parasitic members of the Sarcodina, *Entamoeba histolytica* and its close relatives (pp. 77–85). Certain species of the ciliate *Tetrahymena* are probably today in the early stages of parasitism, since they appear to be able to live equally well inside or outside their (invertebrate) hosts (p. 145). Some ciliates have become highly adapted, physio-

---

[1] Exceptions include the amoebae *Hartmanella* and *Naegleria* (pp. 85–6) and some ciliate endoparasites of invertebrates (Chapter 11).

logically, to a parasitic (or symbiotic) life, e.g. those inhabiting the alimentary tract of ruminants (see pp. 149–51).

Amongst the Zoomastigophorea, parasitism has probably arisen, similarly, independently in several groups: examples are seen amongst the flagellates parasitic in the intestine of vertebrates and invertebrates, some of which have by now become considerably specialized (e.g. the hypermastigid flagellates of insects and termites; see pp. 74–5).

One group of flagellates has specialized *par excellence* in parasitism—the suborder Trypanosomatina of the order Kinetoplastida (Chapter 3). These organisms probably evolved from free-living Zoomastigophorea which became adapted to life within the alimentary canal of primitive invertebrate animals in the pre-Cambrian period over 500 million years ago. From that time they evolved with their hosts, some becoming parasites of nematodes, some of molluscs, others of annelids and insects, as these groups themselves evolved. It seems that the Trypanosomatina underwent an evolutionary 'explosion' in the insects (for reasons which are impossible even to guess at), diverging at first into two main groups: the promastigote group (retaining the primitive anterior position of the flagellar basal body) and the epimastigote stock, which may well have arisen as an adaptation to life in the viscous contents of an insect's gut, the posterior displacement of the flagellar insertion making possible the development of an undulating membrane as an aid to locomotion. These insect trypanosomatids were (and are) transmitted by the ingestion of resistant forms passed out in the host's faeces. When insects adopted the habit of sucking the blood of vertebrates, which they did at least as early as the Oligocene period (about 40 million years ago), their trypanosomatid intestinal parasites were introduced to the possibility of entry into the vertebrate (by contamination of the wound caused by the insect with the latter's faeces, containing resistant forms of the trypanosomatid). Some trypanosomatids were able to take advantage of this opportunity, and thus the genera *Leishmania* and *Trypanosoma* evolved—the former probably from the promastigote stock, and at least most species of the latter from the epimastigote stock (*T. cruzi* may have developed from the promastigotes, and the species of *Trypanosoma* which parasitize aquatic vertebrates and leeches presumably evolved directly from the trypanosomatids of annelids).[1] When certain insects adopted the habit of feeding on plant juices, some of their

---

[1] Some authorities have maintained that the genera *Trypanosoma* and *Leishmania* evolved from flagellates parasitic in the intestine of vertebrates, but the alternative view expressed above is held by the majority of protozoologists today (see Baker, 1965).

promastigote parasites became adapted to life in the plants, thus giving rise to the genus *Phytomonas*: this involved the adoption by the flagellates of transmission via the insect's proboscis, a route also adopted, presumably independently, by the genus *Leishmania* and by some of the species of *Trypanosoma* (the Salivaria) which parasitize mammals.[1]

Amongst the Sporozoa, the other large group of Protozoa which has specialized in parasitism and of which some members have adopted the haematozoan (blood dwelling) habit, evolution has probably proceeded broadly as follows.

The subclass Gregarinia, all of which are parasitic in invertebrates, presumably represents the basal stock of the Sporozoa: from this stock evolution proceeded in a single main line to give rise to those parasites of which the subclass Coccidia are the survivors; and it was in this subclass that the parasitization of vertebrates was first embarked upon by Sporozoa. The Coccidia are divided into three groups (ignoring the Protococcida)—Adeleina, Eimeriina and Haemosporina. Presumably the earliest representatives of the two former groups were, as many still are, monoxenous parasites (i.e. with only one host, invertebrate or vertebrate, playing a part in their life cycle) inhabiting the cells of the host's alimentary canal; but some of those parasitizing vertebrates penetrated deeper into the body of the host, and some became adapted to spending part of their life cycle in the circulating blood cells. This development, associated with the adoption of a heteroxenous life cycle (with the addition of a phase in a blood-sucking invertebrate vector to the original phase within the vertebrate), increased the parasite's chances of finding a new host and was thus of considerable survival value. This sequence of events appears to have occurred at least twice within the Sporozoa: amongst the Adeleina, in the heteroxenous genera of which the sexual individuals (gametocytes) are found in blood cells (sometimes division stages—schizonts—are also found there), and also amongst the eimeriines. In the latter group, most genera are still monoxenous, but in those that have heteroxenous life cycles, it is the transmissive organisms (sporozoites) which are found in blood cells; all other stages occur within various fixed tissue cells of the vertebrate host.

The heteroxenous adeleine type of development, in which sexual forms live in the blood cells and fertilization and post-zygotic

---

[1] It has been suggested that the salivarian trypanosomes evolved, separately from the other insect-borne species, from the leech-transmitted line (for a discussion of this rather esoteric point, see Baker, 1965 and Hoare, 1967).

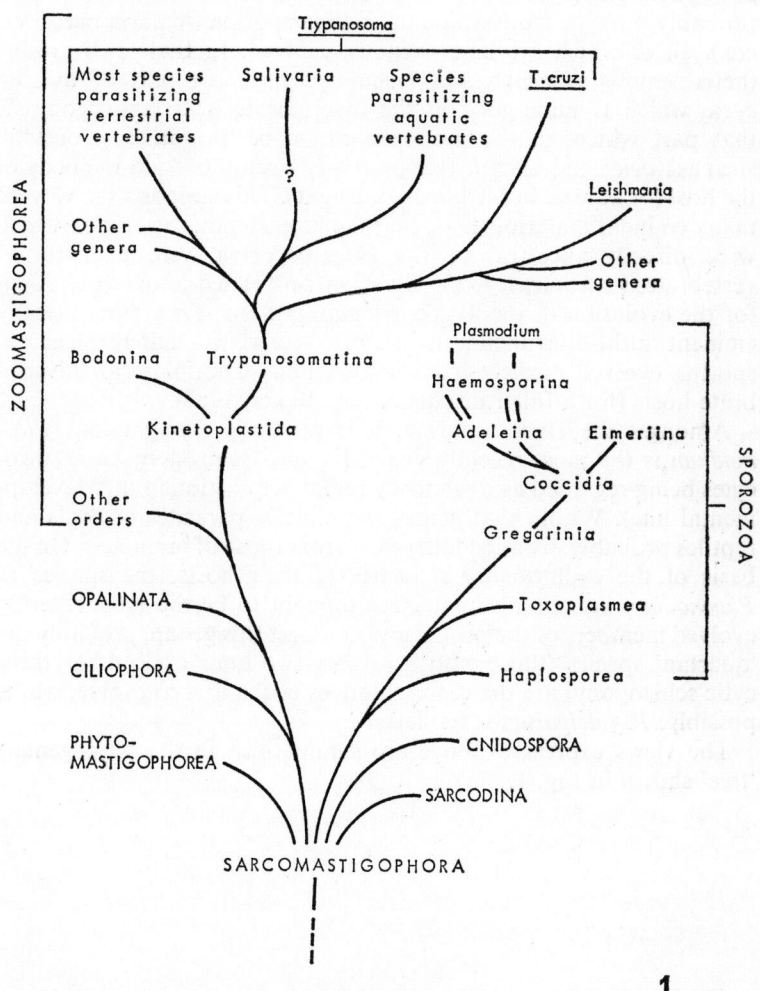

Fig. 1. Simplified phylogenetic 'tree' of the Protozoa, representing their presumed evolutionary development.

multiplication (sporogony) occur in an invertebrate vector, is probably the type from which the Haemosporina (malaria parasites, etc.), all of which are heteroxenous, evolved. In these two groups (heteroxenous Adeleina and Haemosporina), the part of the life cycle which is undergone in the invertebrate host (sporogony) is that part which, in the monoxenous, more 'primitive', Coccidia, such as *Adelea* and *Eimeria* (see pp. 91–7), occurs outside the body of the host (or at least in the lumen of its gut). This supports the view of many eminent malariologists, that the monoxenous Sporozoa which were directly ancestral to the Haemosporina were parasites of vertebrates (in contrast to the situation postulated above (pp. 23–4) for the evolution of the Trypanosomatina). However, other equally eminent authorities incline to the opposite view—that the Haemosporina evolved directly from monoxenous Coccidia with invertebrate hosts (for a fuller discussion, see Baker, 1965).

Amongst the Haemosporina, it is generally thought that *Plasmodium* is the most recently evolved genus (schizogony in erythrocytes being regarded as a relatively recent acquisition in this developmental line). Within that genus, the malaria parasites of birds and reptiles probably diverged fairly early from those of mammals. On the basis of the evolutionary sequence of their hosts, the species of *Plasmodium* infecting primates are thought to be the most recently evolved members of the genus; and amongst this group, probably the 'quartan' species (those with a seventy-two hour cycle of erythrocytic schizogony) are the representatives of the first to evolve, while, possibly, *P. falciparum* is the latest.

The views expressed above are summarized in the phylogenetic 'tree' shown in Fig. 1.

2

ANATOMY AND PHYSIOLOGY OF THE PROTOZOA

There is space in this book for only a very brief discussion of the anatomy and physiology of the Protozoa in general. A good, much fuller account has been published by Dogiel (1965), and rather briefer ones by Manwell (1961) and Mackinnon and Hawes (1961); these should be referred to by those who are interested. Accounts of the finer structure of Protozoa, as revealed by the electron microscope, have been given by Pitelka (1963) and Rudzinska & Vickerman (1968), and these also should be used to supplement the present chapter. Much recent work on the Protozoa which inhabit the blood of their hosts has been collected by Weinman and Ristic (1968).

ANATOMY

As stated in Chapter 1, Protozoa are structurally equivalent to a single metazoan cell: basically, a mass of cytoplasm bounded by some kind of limiting membrane and containing one or more nuclei. In some Protozoa (particularly the Sarcodina), the cytoplasm is divided into two layers—an outer ectoplasm and an inner endoplasm. The cytoplasm (or the endoplasm, in species where this separation occurs) may contain most, if not all, of the organelles found in metazoan cells, including mitochondria, endoplasmic reticulum, ribosomes, Golgi apparatus, bodies resembling lysosomes, fibrils and microtubules of various kinds, centrioles, flagella and cilia. (Not all Protozoa, of course, possess all of these organelles.) Other structures, apparently exclusive to Protozoa, are sometimes present. These are of fewer kinds: examples include the trichocysts of

Paramoecium (a free-living ciliate), various skeletal structures and, perhaps, the contractile vacuoles of ciliates. Some of these organelles will now be considered in a little more detail.

Nucleus

As far as is known, there is no basic structural difference between the nucleus of a protozoon and that of a metazoon. Both are usually spherical or disc-shaped organelles, bounded by a double unit-membrane which is in places continuous with the endoplasmic reticulum. Protozoan nuclei contain deoxyribonucleoprotein and, sometimes at least, ribonucleoprotein which is often concentrated into one or more intranuclear masses called nucleoli. Chromosomes have been reported from the nuclei of some Protozoa: they may well be formed in all, but the small size of the nuclei makes their demonstration difficult. The (haploid) chromosome numbers recorded for most parasitic Protozoa are low: three in several species of *Trypanosoma* (*T. lewisi, T. cruzi* and at least some species of the subgenus *Trypanozoon*) and *Leishmania tropica,* two in most species of *Plasmodium,* three or four in some Cnidospora, and four in some Gregarinia and in *Giardia* spp. The largest numbers recorded for the parasitic Protozoa are in the hypermastigid flagellates living in the gut of termites and certain orthopterous insects, which have from 6 to about 48 chromosomes (Garnham, 1966c; Dogiel, 1965; Grassé, 1952 and 1953). The deoxyribonucleic acid (DNA) content of the nuclei of parasitic Protozoa is also low, in the few organisms for which it has been determined: in *Trypanosoma evansi* the nuclear DNA content is only about 3% of that found in human diploid nuclei (Baker, 1961). The Ciliophora have two nuclei, one sexual nucleus (the micronucleus) and one polyploid asexual one (the macronucleus; see p. 146 below). A few Protozoa have many nuclei (e.g. the Opalinata, p. 75 below), but the majority are uninucleate. Amongst parasitic Protozoa, sexual processes are known only among the Ciliophora, Mastigophora and Sporozoa: in the first, the micronucleus is diploid throughout most of the organism's life history, whilst in some (perhaps all) Sporozoa the nucleus is haploid for all but a brief period of time after fertilization.

Mitochondria

Basically, these are similar in structure to those of most Metazoa, though often the infoldings of the inner membrane (cristae) are tubular rather than the flat, ridge-like structures seen in Metazoa. Probably all parasitic Protozoa have mitochondria at least during some stage of their life cycle, except for anaerobic forms inhabiting the host's alimentary canal, from which they may have been secondarily lost. Some species of malaria parasites (e.g. *Plasmodium berghei*) do not have conventional mitochondria in their asexual stages living in the host's erythrocytes, but possess, instead, organelles composed of concentric membranes which may serve the same purpose (Pitelka, 1963). However, in other stages of their life cycle, typical protozoan mitochondria are present. Certain Mastigophora (the order Kinetoplastida, most of which are parasitic) have an unusually large aggregation of DNA (the kinetoplast) inside their single, large mitochondrion (see p. 40 below).

Secretory organelles

Under this heading are grouped the rough and smooth endoplasmic reticulum and the ribosomes, all of which are present in many if not all of the parasitic and free-living Protozoa which have been examined by electron microscopy. The smooth endoplasmic reticulum is often specialized, as in metazoan cells, into a pile of flattened sacs variously known as a Golgi apparatus or dictyosome. In many Mastigophora the structure originally termed 'parabasal body'[1] seems to be equivalent to a Golgi apparatus.

Fibrils and microtubules

Simple microtubular fibrils have been seen in many Protozoa, often underlying the limiting membrane (see below) of those organisms which have a fairly definite shape—perhaps they serve to maintain this shape. Intranuclear spindle fibrils have also been seen in some species of *Plasmodium,* and may well be of more general occurrence. The development of fibrils seems, amongst the Protozoa, to have reached its maximum in the Ciliophora.

Flagella and cilia

These, as their names imply, are whip-like or hair-like filamentous extensions from the body surface of many Protozoa (and metazoan

---

[1] This must not be confused with the kinetoplast of the order Kinetoplastida, a totally different structure. Unfortunately in older textbooks, including that of Wenyon (1926), this distinction was not always realized and much confusion has resulted.

cells). They are contractile, and beat or wave to and fro in a variety of complex patterns. Cilia, which are usually shorter and more numerous than flagella, normally beat in a synchronous, progressive pattern called metachronal rhythm, aptly described as resembling the appearance of a field of corn waving in the breeze. Electron microscopy has shown that all cilia and flagella (of Protozoa, Metazoa and even plants) have an identical basic structure—a cylinder of nine double microtubules (each resembling a figure 8 in cross section) surrounding a central core of two single fibrils. This '9+2' cylinder (the axoneme) is surrounded by a hollow cylindrical extension of the limiting membrane of the organism (or cell), the sheath. Flagella often (perhaps always) have a band of material within the sheath, parallel to the axoneme, which perhaps has a supporting function related to the fact that flagella are usually considerably longer than cilia. Both flagella and cilia arise from a basal body (kinetosome or basal 'granule')[1] within the cell. The basal body consists essentially of a relatively short extension of the nine outer axoneme fibrils (which may here be triple instead of double), without the two central ones which often appear to originate at a basal plate which marks the junction of axoneme and basal body. In many Mastigophora the flagellar sheath adheres to the organism's limiting membrane and, as the flagellum beats, the membrane is pulled out to form a fin-like undulating membrane. Since both cilia and flagella are primarily locomotory organelles (in Protozoa), the undulating membrane is presumed to help in this function. Certain cilia and flagella secondarily serve as food-gathering organelles, by drawing water currents containing, it is hoped, food particles, into special gullet-like grooves or invaginations of the protozoon's body surface. The propulsive effect of cilia has been likened to that of the oars used in rowing a boat. Some flagella function similarly, while others act as a propeller either pushing or pulling (depending on the direction in which the wave of contraction passes along the flagellum) the organism through its liquid environment. In Protozoa with only one or a few flagella, the latter usually arise anteriorly, though one or more may be recurved to run back along the organism's surface (often forming an undulating membrane, as described above); in some Trypanosomatina, the flagellar origin has moved posteriorly (p. 42).

---

[1] Sometimes called the blepharoplast, but this term—since it was originated for a rather different structure found in the sperm of certain tree ferns—is best discarded from the protozoological vocabulary.

Limiting membrane

Presumably the basic limiting membrane of all cells is the unit membrane (Pitelka, 1963), and this is found as the limiting membrane (plasmalemma) of many Protozoa (e.g. the trypanosomes and amoebae).

In other groups, however (e.g. malaria parasites), a more complicated membrane (or pellicle) has developed, consisting of two or more unit membranes, sometimes apparently with additional material between them—perhaps to give added rigidity. In the Ciliophora, too, a complex pellicle is present, often consisting of an outer unit membrane with a layer of sacs or alveoli beneath it. In some species (e.g. *Paramoecium*), the sacs are inflated, producing a 'sculptured' effect on the pellicular surface, which may be presumed to confer rigidity. The various kinds of protective cyst secreted by many Protozoa may be considered as derivatives of the limiting membrane. Amongst the parasitic Protozoa, cysts are commonly produced by those genera which inhabit their host's alimentary canal and spend part of their life cycle outside the host, while awaiting ingestion by a new host (e.g. parasitic amoebae, and the monoxenous coccidia belonging to the genera *Eimeria, Isospora,* etc.). These cysts serve to protect the organism from desiccation or other damage during this very vulnerable part of its life cycle. Cysts are also produced by some parasitic Protozoa which have developed more efficient means of transmission by vector animals, avoiding the hazards of a 'free' existence, such as the oocysts of *Plasmodium* and the cysts of *Toxoplasma* and *Sarcocystis*. The former may be regarded as an evolutionary legacy from their ancestors (see Chapter 1), while the function of the latter is probably still protective—but against the host's defences rather than against a hostile environment outside the host.

Skeletal structures

Many free-living Protozoa have elaborate exo-skeletons (e.g. radiolaria, foraminifera) but these are absent from parasitic forms (unless cysts are considered as exo-skeletal structures). Perhaps the tough wall which surrounds the spore of the Cnidospora may legitimately be regarded as exo-skeletal.

Endo-skeletal structures are found in some parasitic Protozoa, particularly the Mastigophora (e.g. the axostyle of *Trichomonas*), but these, too, are relatively rare. As mentioned above, the sub-pellicular microtubules of trypanosomatids and malaria parasites may be endo-skeletal in function.

Contractile vacuoles

As the primary function of these organelles is to remove unwanted water entering the organism by osmosis or during feeding, they are rarely found in parasitic Protozoa, which usually inhabit an isotonic environment. However, they are seen in parasitic ciliates—and the amoebae *Hartmanella* and *Naegleria* which sometimes live parasitically in man (p. 85 below). A contractile vacuole consists of a rhythmically pulsating vesicle, which is fed by a system of radial canals and opens to the outside through a small pore in the organism's limiting membrane. A complicated arrangement of 'valves' ensures that liquid flows in only one direction—from the canals into the vacuole and then, when the latter contracts, out through the pore. Further details of these organelles can be obtained from a recent review by Kitching (1967).

PHYSIOLOGY
Locomotion

The locomotion of Protozoa has been reviewed by Jahn and Bovee (1967), and that of the parasitic species in particular has been discussed by Garnham (1966b). Basically there are three known methods by which Protozoa move: (1) amoeboid movement, (2) flagellar and ciliar movement, and (3) gregarine movement. These will be discussed briefly one by one.

(1) Amoeboid movement. Not surprisingly, this is the method by which amoebae move. It is characteristic of the whole superclass Sarcodina and is also used by some Mastigophora and a few Sporozoa (at certain stages of the life cycle), but not by any of the Ciliophora (presumably the complexity of the pellicle of this latter group precludes it). Amoeboid movement consists of the temporary extension of the body into one or more processes, called pseudopodia ('false feet'), in the direction in which movement is occurring; the rest of the body is then drawn up into the pseudopodium, and the whole procedure repeated. Pseudopodia may be of different kinds—lobopodia (broad and blunt in shape), filopodia (long and thin), reticulopodia (thin and branching, forming a network) and axopodia

(long and thin, with a central supporting rod). The classification of the class Rhizopodea is based largely on the type of pseudopodium present (see Chapter 1). All the parasitic amoebae have lobopodia. These appear first as a protrusion of the ectoplasm (the outer, clear layer of cytoplasm) at the advancing end of the organism; as this protrusion extends, the endoplasm begins to flow into it, and gradually the whole organism 'catches up', as it were, with the pseudopodium.

The precise mechanism involved in the production of pseudopodia is still not completely understood, but the most popular theory at present is that it depends upon active contraction of the ectoplasmic 'tube' at the hind end of the body: thus the endoplasm is squeezed forwards into the expanding pseudopodium, while the ectoplasm forming the sides of the tube remains stationary. This process must involve continual transformation of ectoplasm to endoplasm and breakdown or absorption of the pellicle at the hind end, and their re-formation (from endoplasm) at the front. Amoebae thus have a definite polarity—the hind end, or 'uroid', being a relatively constant feature. Whether this polarity remains unchanged throughout the life of the amoeba (i.e. the period from one cell-division to the next) is unknown. There is evidence that adenosine-triphosphate (as an energy carrier) and proteins resembling actomyosin play a part in amoeboid movement, as indeed they do in metazoan muscular contraction and in the beating of cilia and flagella (see Jahn and Bovee, 1967).

(2) Flagellar and ciliar movement. The structure of flagella and cilia, and the main types of beat characteristic of each, have been described above (pp. 29–30). For a fuller treatment see the review by Jahn and Bovee (1967).

The submicroscopic mechanism producing the beat is not known: it is presumed to involve contraction of some or all of the axonemal fibrils. It is interesting that, here again, the mechanism depends upon the presence of adenosinetriphosphate as an energy 'carrier', and that proteins similar to actomyosin have been extracted from cilia, so that the molecular mechanism involved may well be similar to that found in metazoan muscle cells: indeed, it is likely that all forms of cytoplasmic movement are dependent on the same basic mechanism (Jahn and Bovee, 1967).

The co-ordinating mechanism responsible for synchronizing the metachronal rhythm of beating cilia (see p. 30) is not known: opinion is divided as to whether it depends upon impulse conduction along some of the fibrils which appear to link the ciliary basal bodies, or whether it is a purely mechanical effect transmitted from cilium to cilium by pressure waves in the surrounding medium.

Flagellar and ciliar movement is, of course, typical of the Mastigophora and Ciliophora. It is also seen in the Opalinata and a few Sarcodina (at certain stages of the life cycle). Amongst the Sporozoa, only microgametes of the subclass Coccidia and of some members of the subclass Gregarinia are flagellate.

(3) Gregarine movement. This is typical of the Sporozoa at certain stages of their life cycles. It is probably the least understood of all the types of protozoan locomotion. An organism moving in this way, when seen under a light microscope, seems merely to glide along, without using any special organelles. Gregarine movement may be produced by small-scale contraction or 'rippling' of the pellicle, moving the organism in a way similar to that in which a slug moves, though other hypotheses have been suggested (Jahn and Bovee, 1967).

Movement of this kind (or believed to be of this kind) is shown by many Gregarinia (hence its name), by gametocytes of eimeriine haemogregarines, by ookinetes of the Haemosporina, by sporozoites of many of the Sporozoa, and by the Toxoplasmea.

Nutrition

Nutrition may be either holophytic (plant-like or synthetic, based on chlorophyll) or holozoic (animal-like or analytic, based on predation). Those free-living Protozoa which possess chloroplasts have holophytic nutrition while the rest have the holozoic variety. The latter consists essentially of the ingestion of complex molecules of proteins, carbohydrates, fats, etc., in the form of other animals or plants (either whole or in pieces), breaking these molecules down into simpler units (amino acids, simple sugars, fatty acids) and rebuilding them into the animal's own proteins, carbohydrates and fats.

When holozoic Protozoa obtain their food by actively ingesting large particles (whole cells, or large chunks of them), the feeding process is said to be phagotrophic. This, and sometimes even holozoic nutrition itself, may be contrasted with what is termed saprozoic nutrition, in which simpler organic molecules (amino acids, simple sugars and fatty acids), produced as a result of decay or (as in the case of parasites) the host's digestive processes, are absorbed by the protozoon in solution by diffusion through the pellicle. However, the discovery (by electron microscopy) that phagotrophy can occur on a sub-(light)-microscopical scale (in *Plasmodium*; see p. 35

below), and the fact that the trypanosomes, which had always been thought to be saprozoic feeders, are now known to feed by a process called pinocytosis[1] which really differs only in degree from phagotrophy, has led to the apparent distinction between holozoic and saprozoic nutrition becoming less clear.

Parasitic Protozoa show a trend towards dietary specialization, presumably due to the production of fewer digestive enzymes, compared with their free-living relatives. This is reflected in the fact that they are usually more difficult to cultivate *in vitro,* in non-living media, than are non-parasitic forms. At one end of the scale are the parasitic amoebae and flagellates, almost, if not quite, as easy to cultivate as their free-living relatives; at the other end are the obligate intracellular forms, such as *Toxoplasma* and (for most of its life cycle) *Plasmodium,* which are either impossible or very difficult even to keep alive, let alone to grow, outside a suitable living cell (see Taylor and Baker, 1968).

The food-gathering mechanisms used by the parasitic Protozoa are generally the same as those of their non-parasitic relatives. Amongst the Zoomastigophorea, many of the forms inhabiting the host's intestine have a distinct mouth or cytostome through which quite large food particles are ingested (phagotrophy). The cytostome is at the base of a groove or cytopharynx containing a flagellum, which produces water currents to draw food particles down the groove. A similar mechanism is employed by many parasitic Ciliophora. The Zoomastigophorea which inhabit the host's blood or tissue fluids (the trypanosomes) feed rather similarly, but do not have a single, relatively large mouth. Instead, small (submicroscopic) droplets of the host's plasma are ingested by pinocytosis (see footnote below) or through a cytostome within the flagellar invagination (a deep intucking through which the proximal part of the flagellum passes).

Probably all Sporozoa feed by phagotrophy, at least at certain stages of their life cycle, through one or more cytostomes (small cylindrical depressions) in the thick pellicle (in Sporozoa the pellicle consists of at least two unit membranes, and only at the base of the cytostome is the cytoplasm bounded by a single unit membrane).

Parasitic Rhizopodea of the genus *Entamoeba* feed by phagotrophy, taking in food particles at the hind end (uroid), where the pellicle of

---

[1] Pinocytosis, also seen in metazoan cells, consists of the ingestion of minute droplets of the surrounding fluid by their incorporation into tiny vacuoles formed by invagination of the surface (unit) membrane of the cell, followed by 'pinching off' the vacuole, and the inward passage of the vacuole into the cytoplasm.

the parasite seems to be sticky: particles adhere to it and become drawn into the amoeba's cytoplasm when the pellicle itself disintegrates and becomes drawn in at this region during the process of locomotion (see p. 33 above). This is rather different from the engulfment of food particles by pseudopodia which is the way in which most free-living Rhizopodea catch their food. It is uncertain which method is adopted by the other genera of parasitic amoebae.

It is not known whether simple diffusion plays an important part in the ingestion of nutrients by parasitic Protozoa, in addition to the processes outlined above, but it is perhaps unlikely to do so (see pp. 34–5 above).

The digestive processes by which the Protozoa break down the nutrients which they ingest are little known, but are probably similar, at least in outline, to those used by the Metazoa (see Lwoff, 1951 and Hutner and Lwoff, 1955).

Respiration

The purpose of feeding is to obtain raw materials for growth and repair, and also to obtain fuel from which the energy required by the cells can be released. The processes by which energy is released are called respiration. Respiration may be aerobic (requiring oxygen) or anaerobic. Examples of both are found among the parasitic Protozoa (see Danforth, 1967). Generally, those which live in blood or in tissue cells are aerobic, obtaining their oxygen from the host's blood or tissue fluid, while those living in the alimentary canal where the oxygen tension is low, are anaerobic. As far as is known, aerobic respiration in almost all parasitic Protozoa follows the same general pattern as it does in metazoan cells, being based upon the oxidation of glucose to carbon dioxide and water, via the Embden–Meyerhof pathway, Krebs's tricarboxylic acid cycle and a cytochrome system, the two latter systems being—again as in metazoan cells—associated with the mitochondria. One interesting exception has, however, been found. The stages of *Trypanosoma brucei* (and its subspecies) which multiply in the blood of the mammalian host (but not those which develop in the insect vector) have adopted an apparently unique non-mitochondrial system of aerobic respiration. The initial stages of glycolysis (Embden–Meyerhof pathway) are the same, but glucose is degraded only as far as pyruvic acid, which is excreted. The subsequent stages of oxidation, to carbon dioxide and water, and oxidative phosphorylation (involving Krebs's cycle and cytochrome system), both of which normally take place in mitochondria, do not occur.

Instead, terminal respiration is mediated by an L-α-glycerophosphate oxidase–L-α-glycerophosphate dehydrogenase system, located possibly in small membrane-bound vesicles which are found throughout the cytoplasm (Vickerman, 1965).

The energy-releasing fermentations adopted by anaerobic parasitic Protozoa are discussed in the review by Danforth (1967).

Excretion

The excretion of soluble waste products from parasitic (and other) Protozoa usually occurs by diffusion. Protozoa which have contractile vacuoles doubtless remove some soluble wastes via these organelles, but their main function is osmo-regulatory.

Insoluble matter is ejected from the food vacuoles of Ciliophora through a small pore (cytopyge) in the pellicle. Rhizopodea behave similarly but, since they are bounded by only a single unit membrane, do not seem to require a special pore. Malaria parasites convert the insoluble residue (including iron) of the haemoglobin on which they feed to brown or black pigment (haemozoin or malaria pigment) and parcel it up in vacuoles within their cytoplasm; it is then left behind during the next division process.

Asexual reproduction

This can occur in one of at least five ways.

(1) Binary fission. This is the simple division of one individual into two, and is the commonest form of asexual multiplication. The organism's nucleus first divides by mitosis. The details of nuclear division have been elucidated completely in very few Protozoa, partly because of the small size of the nuclei. In some, but not all, Protozoa mitosis occurs within the intact nuclear membrane (differing from the process in Metazoa): chromosomes, spindle fibrils, etc., have been seen during a few protozoan mitoses and are probably generally if not universally formed. There is doubt as to precisely how the polyploid macronuclei of the Ciliophora divide.

Following nuclear division (karyokinesis), the organism's cytoplasm divides (cytokinesis): in most groups (Rhizopodea, Ciliophora and Sporozoa) this appears to result from the development and deepening of a furrow around the organism in the plane of fission, which in Ciliophora and those Sporozoa which undergo binary fission is equatorial or transverse (division being said to be homothetogenic in these groups). In the Mastigophora, the plane of fission is meridional or longitudinal (division being said to be symmetrogenic), and the separation of the two 'daughter' organisms begins anteriorly and proceeds steadily back. Some organelles divide at the time of division (e.g. nucleus, mitochondria presumably), while others have to be formed anew by one of the 'daughters' (e.g. flagella, cytostomes, and the ciliary mouth complexes of Ciliophora).

(2) Multiple fission. This is a variant of binary fission in which a subsequent division commences before the earlier one was completed. It is seen particularly amongst Eugregarinida (Sporozoa), resulting in long chains of individuals, and in Trypanosomatina (Zoomastigophorea), where it may result in the formation of spheres or rosettes of organisms, all still attached at their hind ends.

(3) Budding. In this type of division a new, small individual develops as a bud on the surface of the old, one of the products of nuclear division entering the bud; finally the bud breaks off and grows to full size. This method is common among certain Ciliophora but rare (possibly absent) among parasitic Protozoa.

(4) Schizogony. Originally regarded as a form of multiple fission, electron microscopy has shown schizogony to be rather a type of multiple budding. After two or more nuclear divisions, the 'daughter' nuclei move to the periphery of the parent organism (schizont) and 'daughter' individuals or merozoites develop as buds, one related to each nucleus. Eventually the merozoites (into each of which one nucleus has entered) break off; the schizont is destroyed in the process, all that remains of it being a residual body of cytoplasm (and, in the case of erythrocytic schizonts of *Plasmodium,* the malarial pigment). It is characteristic of schizogony that bud formation does not begin until after all nuclear divisions are complete. The number of merozoites produced ranges from four to many thousands. In some very large schizonts the area available for the production of merozoites is increased by complex invaginations of the surface, and this may sometimes result in the complete separation of parts of the schizont, which are then called cytomeres. Schizogony is found only in the Sporozoa (and, perhaps, in the Microsporida; see p. 142).

It is interesting to note that—at least in the genus *Plasmodium* (Coccidia, Eucoccida, Haemosporina)—the way in which the sporozoites are formed during sporogony (see p. 104 below) is very similar to schizogony, and probably the two processes should be regarded as the same.

(5) Endodyogeny. This rather rare type of asexual reproduction, so far recorded only in the Toxoplasmea (Sporozoa), consists of the

Anatomy and physiology of the Protozoa

development of two 'daughter' individuals within a single parent, which is destroyed in the process. Although sometimes called 'internal budding', endodyogeny may *perhaps* be regarded as a special kind of schizogony.

Sexual reproduction

Among parasitic Protozoa, sexual reproduction is known to occur only in the Sporozoa, Opalinata, Hypermastigida (Zoomastigophorea parasitic in the gut of termites and insects) and Ciliophora. The latter have a very specialized form of sexual reproduction (conjugation), which is described in Chapter 11 below. In the Sporozoa, sexual reproduction (see p. 94) consists basically of the fusion (copulation) of two sexual individuals, male and female, usually anisogametes, as they differ in size and in the fact that the male gametes (except in certain Gregarinia) bear one or more flagella, while the females do not.

Sexual reproduction in the Opalinata (p. 75) occurs by the fusion of two multiflagellate anisogametes (which differ slightly in size). These gametes emerge from cysts which are produced only during the host's breeding season, apparently under the influence of the onset of sexuality in the host itself (which is usually an amphibian). The cysts are therefore available to infect the young tadpoles, in which fertilization occurs.

The occurrence of sexual reproduction among the hypermastigid flagellates (p. 74) is controlled by the production of moulting hormone by the host. It consists of the fusion of two flagellate gametes which may be similar or dissimilar, i.e. isogamous or anisogamous and which in some species are indistinguishable from the trophozoites.

Reduction division of the nucleus (meiosis) occurs at varying stages in the life cycle of those Protozoa which undergo sexual reproduction. In the Sporozoa, and in some hypermastigid flagellates, it occurs at the first nuclear division of the zygote; in other hypermastigids, in the Ciliophora and in the Opalinata, it occurs during the production of the gametes (as it does in *Homo sapiens*). Thus the former organisms are haploid throughout most of their life cycle, and the latter are diploid. The process of meiosis in Protozoa is, as far as is known, very similar to that in the metazoa.

A more detailed account of sexual reproduction among Protozoa is given by Grell (1967).

3

TRYPANOSOMES AND RELATED PARASITES

These organisms belong to the class Zoomastigophorea, order Kinetoplastida (see Chapter 1), so named because its members possess an organelle called a kinetoplast. This is seen by light microscopy as a small, usually round or oval body, situated near the base of the flagellum, which stains similarly to a nucleus. For many years its true nature has puzzled protozoologists, some of whom at one time thought it to be a second nucleus. The electron microscope has recently (since 1960) revealed that the kinetoplast is a mass of DNA contained within a very large mitochondrion, a most exciting discovery because, until then, it was almost an axiom of cell biology that DNA occurred only in nuclei. It has subsequently been recognized that DNA occurs in mitochondria of many other organisms, and that what is special about the Kinetoplastida is only that they have more of it there than do other animal cells (Vickerman, 1965). No sexual development is known for any member of the order. There are two suborders within the Kinetoplastida, the Bodonina and the Trypanosomatina, the latter being the more important from the parasitologist's viewpoint.

Suborder 1. Bodonina

Many of these organisms are free-living. They can be distinguished from the Trypanosomatina by having more than one flagellum (usually two). Most of the parasitic forms live in the intestines of vertebrates and invertebrates, but some species of the genus *Cryptobia* (Fig. 2) inhabit the blood of fish and these are often called trypanoplasms, since they were once grouped in a separate genus, *Trypano-*

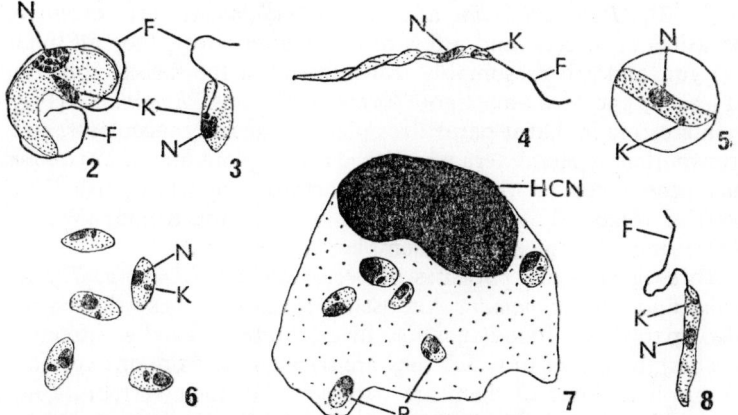

Fig. 2. *Cryptobia keysselitzi* in the blood of a fish (tench). ×1,500. (From a slide kindly supplied by Dr E. A. Needham.)
    3. *Crithidia* sp. from an insect. ×1,500
    4. *Phytomonas* sp. from Euphorbiaceae in Tanzania; promastigote form. ×1,500.
    5. *Endotrypanum* sp. in erythrocyte of a sloth. ×1,500.
    6. *Leishmania brasiliensis*; amastigote forms liberated from the host cell in a smear from a lesion in man. ×1,500.
    7. *L. tropica*; amastigote forms in a macrophage from a smear made from an oriental sore. ×1,500.
    8. *L. tropica mexicana*; promastigote form from a culture. ×1,500.
Abbreviations: F – flagellum; HCN – host cell's nucleus; K – kinetoplast; N – nucleus; P – parasite.

plasma. They are not usually pathogenic. Trypanoplasms are transmitted by leeches, in which they multiply in the gut. Eventually they migrate forwards to the proboscis sheath, and some are injected into the blood of the next fish on which the leech feeds.

Suborder 2. Trypanosomatina

All members of this suborder are parasitic and several are important pathogens. All are elongate, slender Protozoa at least at some stage of their life cycle, with a single nucleus and a kinetoplast situated near the origin of the single, anterior flagellum, by means of which they swim actively. Different forms are recognized, depending upon the position in the body of the kinetoplast and basal body and the course taken by the flagellum (Hoare & Wallace, 1966). Some genera exist for part of their life cycle as non-flagellate or amastigote individuals. The genera *Leptomonas, Herpetomonas, Crithidia*

(Fig. 3), *Blastocrithidia* and *Rhynchoidomonas* are exclusively parasites of insects (and a few other invertebrates); they all inhabit the gut, and are presumably transmitted via the faeces, sometimes at least as encysted amastigote forms. The genus *Phytomonas* (Fig. 4) is interesting in that it parasitizes plants, especially succulents, and is transmitted by Hemiptera which feed on the plant juices. *Phytomonas* has been recorded from various parts of the world, usually the warmer regions but by no means always in the tropics (Wenyon, 1926); morphologically, it resembles *Leptomonas*.

Three genera are parasites of vertebrates: *Leishmania, Trypanosoma* and *Endotrypanum*. Almost all species of these three genera also parasitize blood-sucking invertebrates (usually insects or leeches), by means of which they are transmitted from one vertebrate to another. Some of them are pathogenic to their vertebrate hosts but, with one possible exception *(T. rangeli),* there is no evidence that any harms its invertebrate host. *Endotrypanum* (Fig. 5) is unique among the Trypanosomatidae in that it lives inside the erythrocytes of its hosts (sloths of South and Central America); the parasites are either epi- or promastigote (see below), and are perhaps transmitted by sandflies (see p. 43). The full life cycle is unknown.

Leishmania exists in two forms—amastigote (i.e. rounded, non-flagellate) individuals (Fig. 6) in its vertebrate host and elongated, flagellate organisms in the invertebrate. The flagellate form is of the type known as promastigote, in which the basal body and kinetoplast are close to the anterior end and the flagellum emerges through a short intucking of the body surface, the 'reservoir' or 'flagellar pocket' (Fig. 8). (*Leptomonas* and *Phytomonas* also exist in these two forms.)

The genus *Trypanosoma* is characterized by the fact that, at some stage of their life history, all its species occur as trypomastigote forms (Fig. 9)—slender, flagellate individuals in which the kinetoplast and basal body are near the posterior end, and the flagellum emerges through a short 'pocket', running anterolaterally. The flagellum then passes forwards along the surface of the organism's body, to which it is apparently attached in such a way that, as the flagellum undulates, it draws out a fin-like expansion of the body to form the undulating membrane. Some species of *Trypanosoma* also exist, at different stages of their life cycle, as amastigotes and promastigotes, as well as in a fourth form, the epimastigote (Fig. 16). In this last form, the kinetoplast and basal body lie close to, usually beside, the nucleus; the flagellum emerges, and is attached to the

Trypanosomes and related parasites

body to form an undulating membrane, in the same way as in the trypomastigote form. Other species of *Trypanosoma* may lack the amastigote and promastigote stages, while others exist only as trypomastigotes. It is possible that the amastigote–promastigote type represents the ancestral form of the subfamily and that the epi- and trypomastigote stages are subsequent evolutionary additions to the primitive cycle (see Chapter 1); thus the life cycle of *Trypanosoma* can be regarded as an ontogenic recapitulation of phylogeny, from which some of the more recently evolved species have omitted one or more of the 'ancestral' stages.

Genus *Leishmania* Ross, 1903

Kinetoplastida which parasitize lymphoid–macrophage cells of vertebrates (mammals and reptiles) and the gut of sand-flies (Diptera, Psychodidae, Phlebotominae); they exist as amastigote forms in the former, and as promastigotes in the latter.

Morphology and life cycle

The morphology and life cycle are similar in those species for which they are known. In the vertebrate host the parasites exist solely as amastigote forms, usually 2–4 μ in diameter, often called 'Leishman–Donovan' or 'L.D.', bodies after the names of their first observers (Fig. 6). These are ingested by the macrophages as part of the latter's phagocytic activity, but instead of being destroyed, the parasites thrive and multiply by binary fission within the macrophages (Fig. 7). Probably, when the infected cell divides, the parasites are shared between the two daughter cells. If an infected macrophage dies, the liberated parasites are presumably ingested by other macrophages. Thus more cells become infected. Infected macrophages in the blood or skin are ingested by sand-flies (*Phlebotomus* spp.) when the insects are feeding (only the females feed on blood). In the mid-gut of the sand-fly, the parasites emerge from the macrophages, transform into promastigote forms (Fig. 8) by elongating and developing a flagellum, and commence multiplying by binary fission. The flagellates spread forwards through the insect's gut, many of them becoming attached to the mucosa by their flagella. They become established in the foregut, and by their multiplication eventually block the cavity of the proventriculus, pharynx and proboscis. At this stage, about ten days after it first ingested parasites, when the insect again attempts to feed (which it usually does about every five days), its efforts to pump saliva down the blocked proboscis and to draw up blood result in the injection of many flagellates into the host's skin. Here they are phagocytosed by macrophages and, after reverting to the amastigote form, they commence dividing. Transmission may also result from the crushing of infected sand-flies while they are feeding, their gut contents entering the vertebrate through the puncture made by the proboscis. Blood transfusion from an infected donor has also occasionally transmitted the infection.

Important species. (1) *Leishmania donovani* causes visceral leishmaniasis or kala-azar of man throughout the warmer parts of Asia, the Mediterranean coasts, North and East Africa and South America. In some areas (the Mediterranean, China, central Asia, South America) dogs are often infected and serve as reservoirs of the parasite; in East Africa and Sudan this role is played by wild rodents, while in India man is the only vertebrate host. Parasitized macrophages occur in all tissues, including the blood; the disease is slow but, unless treated, is usually fatal. Treatment is with compounds of antimony, and is often, but not always, effective. It is sometimes followed by a nodular eruption of the skin, which may be allergic in origin, called post-kala-azar dermal leishmanoid; parasites are found in the nodules.

(2) *L. tropica* causes cutaneous leishmaniasis or oriental sore, which has a similar but slightly more restricted distribution than kala-azar (e.g. it is absent from south India and East Africa). Dogs and wild rodents are reservoir hosts. Parasites are restricted to ulcers in the skin—usually only one or a few on each patient. The disease is mild, usually ending spontaneously. There are many local varieties of cutaneous leishmaniasis, and some authorities consider them to be caused by different species or subspecies of the parasite.

(3) *L. braziliensis* is sometimes regarded as a subspecies of *L. tropica*. It occurs in Brazil, and may cause a very severe human disease—mucocutaneous leishmaniasis or espundia; the mucous membranes of the nose, mouth and pharynx become infected and ultimately destroyed. Spontaneous recovery is rare and treatment, with antimony compounds, is difficult. It is not known whether there are vertebrate hosts other than man.

Diagnosis

Human leishmaniasis may be diagnosed by examining smears of material obtained by puncturing or scraping suspected lesions;

the smears are stained with Giemsa's stain in the same way as a thin blood film (see Chapter 13) and parasites, if present, can be seen. A more sensitive method is to inoculate material, obtained as described above, into blood-agar cultures (see Chapter 13) and see whether promastigotes develop; for this method, the material from the lesions must be obtained aseptically.

Other species. Apart from infections of *L. donovani* and *L. tropica* in dogs, and *L. tropica* sometimes in cats, there is no leishmaniasis of domestic animals. Wild rodents and lizards may be infected with species of *Leishmania,* which are morphologically similar to those of man. *L. enriettii,* a parasite of South American guinea-pigs, has unusually large amastigotes—about 3×6 μ.

Genus *Trypanosoma* Gruby, 1843

Kinetoplastida parasitizing the blood and sometimes other tissues of vertebrates and, usually, the gut of blood-sucking invertebrates. They exist as trypomastigote forms for at least part of the life cycle in both hosts.

Trypanosomes of non-mammals. Species of *Trypanosoma* are common parasites of fishes (fresh-water and marine), amphibia, reptiles and birds. Usually these species are relatively large organisms (about 50–100 μ long) and are very scanty in the blood of their hosts. All, as far as is known, are non-pathogenic. The vectors (transmitting hosts, see p. 13) of the trypanosomes of terrestrial vertebrates are blood-sucking arthropods (usually insects, occasionally mites); the trypanosomes of aquatic forms are transmitted by leeches (Annelida, Hirudinea). In the vector's intestine they multiply firstly as epimastigote forms and then change into small trypomastigote forms ('metacyclic forms' or metatrypanosomes). The latter are capable of reinfecting the vertebrate host; they usually develop in the vector's hind gut and are introduced to the vertebrate either by faecal contamination of the wound produced by the vector when it feeds, or by the vertebrate ingesting infected vectors and crushing them in its mouth. The species transmitted by leeches, however, often complete their development in the vector's fore-gut and so are injected with the leech's saliva into the next vertebrate on which it feeds. Much remains unknown about this group of trypanosomes. British wild birds (especially Corvidae) are commonly infected (Fig. 9), as are carp, goldfish and other fish as well as frogs. The parasite of the frog, *T. rotatorium,* appears normal in the tadpole but, after the

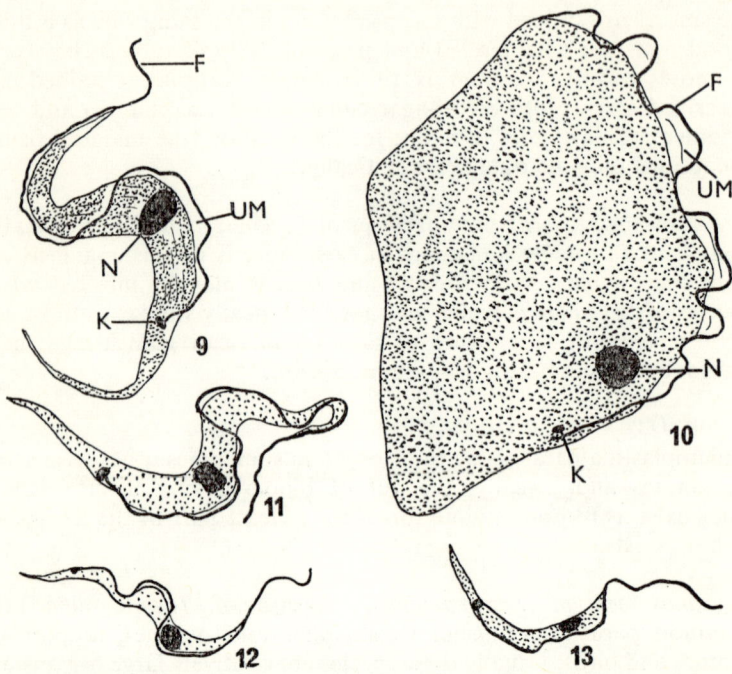

Fig. 9. *Trypanosoma avium*; trypomastigote form in the blood of a canary. ×1,500.
    10. *T. rotatorium* in the blood of a Liberian frog. ×1,500.
    11. *T. theileri* in the blood of a Nigerian ox. ×1,500. (From a slide kindly supplied by Mr R. Killick-Kendrick.)
    12. *T. rangeli* in mouse blood. ×1,500.
    13. *T. lewisi* in rat blood. ×1,500.

Abbreviations: F – flagellum; K – kinetoplast; N – nucleus; UM – undulating membrane.

latter's metamorphosis into the adult frog, becomes enlarged and 'leaf-like' in shape (Fig. 10).

Trypanosomes of mammals. All the important pathogens are included in this group, which has been divided by Hoare (1966) into two sections (Stercoraria and Salivaria), each containing several subgenera. The chief morphological characteristics of the more important species of both sections are summarized in Table 1, and a simplified 'key' for their identification in stained blood films is given in Table 2.

TABLE 1

MAIN MORPHOLOGICAL FEATURES OF SOME MAMMALIAN TRYPANOSOMES AS SEEN IN BLOOD OR TISSUE FLUIDS OF THE VERTEBRATE HOST

Subgenus	Species	Average length	Average breadth	Kinetoplast Size	Kinetoplast Position	Posterior end	Free flagellum	Position of nucleus
Megatrypanum	T. theileri	60–100μ	2–3μ	Medium	Not terminal	Pointed (long)	Long	Central
	T. melophagium	40–60μ	2–3μ	Medium	Not terminal	Pointed (long)	Long	Central
Herpetosoma	T. rangeli	25–35μ	2μ	Large, rod-like	Not terminal	Pointed (long)	Long	Slightly anterior
	T. lewisi	21–36μ	2μ	Large, rod-like	Not terminal	Pointed (long)	Long	Slightly anterior
Schizotrypanum	T. cruzi	15–24μ	1·5μ	Large	Sub-terminal	Pointed (short)	Long	Central
Dutonella	T. vivax vivax	21–26μ	1·5–2μ	Large	Terminal	Blunt, rounded	Long	Central
	T. v. viennei	21–26μ	1·5–2μ	Large	Terminal	Blunt, rounded	Long	Central
	T. uniforme	14–17μ	1·5μ	Large	Terminal	Blunt, rounded	Long	Central
Nannomonas	T. congolense	12–18μ	1–2μ	Medium	Sub-terminal, marginal	Usually blunt	Usually absent	Central
	T. simiae	17–18μ	1–2μ	Medium	Sub-terminal, marginal	Usually blunt	Usually absent	Central
Pycnomonas	T. suis	15μ	3·5μ	Small	Sub-terminal	Pointed (very short)	Short	Central
Trypanozoon	T. brucei subspecies Slender forms	30μ	1·5μ	Small	Sub-terminal	Blunt, often truncate	Long	Central
	Stumpy forms	18μ	3·5μ	Small	Sub-terminal	Blunt, rounded	Usually absent	Central (except in posteronuclear form)
	T. evansi evansi	20–28μ	1·5μ	Small	Sub-terminal	Blunt or truncate	Long	Central
	T. e. equiperdum	20–28μ	1·5μ	Small	Sub-terminal	Blunt or truncate	Long	Central
	T. equinum	20–28μ	1·5μ	Absent	—	Blunt or truncate	Long	Central

TABLE 2

A SIMPLIFIED KEY TO THE MORE IMPORTANT MAMMALIAN TRYPANOSOMES IN STAINED THIN BLOOD FILMS

1.	Trypanosomes more than 40 μ long	*T. theileri*, etc.
	Trypanosomes less than 40 μ long	2
2.	All individuals with free flagellum	3
	Some or all individuals without free flagellum	11
3.	Kinetoplast not terminal, medium-sized to small (or absent)	4
	Kinetoplast terminal, large	9
4.	Posterior end sharply pointed	5
	Posterior end bluntly pointed or rounded	10
5.	Kinetoplast medium-sized or large	6
	Kinetoplast small	8
6.	Trypanosomes curved (C-shaped); kinetoplast large, round	*T. cruzi**
	Not as above	7
7.	Kinetoplast medium-sized, round	*T. rangeli**
	Kinetoplast large, rod-shaped	*T. lewisi*
8.	Long, slender trypanosomes	*T. rangeli**
	Short, broad trypanosomes (rare Central African species)	*T. suis*
9.	Trypanosomes more than 20 μ long	*T. vivax*
	Trypanosomes less than 18 μ long	*T. uniforme*
10.	Kinetoplast present, small	*T. e. evansi*, *T. e. equiperdum* and old laboratory strains of *T. brucei* sspp.
	Kinetoplast absent	*T. equinum*
11.	All individuals without free flagellum	*T. congolense*
	Some individuals have, or appear to have, free flagellum	12
12.	Differentiated into long slender forms with free flagellum, short stumpy forms without free flagellum, and intermediates; kinetoplast not markedly marginal in position	13
	Not as above; kinetoplast marginal	*T. simiae*
13.	Parasitaemia in laboratory rodents high; posteronuclear forms common	*T. b. brucei* and *T. b. rhodesiense**
	Parasitaemia in laboratory rodents low; posteronuclear forms rare	*T. b. gambiense**

Notes 1. In practice, the species of animal and the geographical location from which the trypanosomes were isolated are usually valuable aids to identification.

2. In the key, species or sub-species which can infect man are marked with an asterisk(*).

3. This key gives only certain differential characters of the groups concerned: it is not a full description of them, *nor does it represent their phylogenetic relationships.*

Section A. Stercoraria

All members of this section have two hosts, a mammal and an insect vector. The insect ingests trypomastigote forms when it feeds on the blood of an infected mammal; these forms quickly change into the epimastigote stage, chiefly by a forward movement of the kinetoplast and flagellar basal body (or kinetosome). The epimastigote forms then divide repeatedly, by longitudinal binary fission, in the insect's mid-gut, and gradually migrate backwards to colonize the hind-gut. Finally, metatrypanosomes develop in the hind-gut (except for *T. rangeli*; see below). They probably do not divide until after they have entered the mammalian host. The fact that they develop in the hind-gut of the vector is one of the main features separating the Stercoraria from the Salivaria. Metatrypanosomes may be passed out in the vector insect's faeces, or they may be liberated from an infected insect if the latter is crushed by the mammal on which it is feeding. They can then enter the mammal's blood-stream in various ways: they may penetrate the puncture produced by the insect's proboscis; they may be rubbed into lesions produced by scratching the skin in response to the irritation of the insect's bite; they may penetrate the mucous membrane[1] of the mouth if the mammal either licks up insect faeces, or crushes whole insects in its mouth, during its cleaning operations. All these methods are examples of contaminative transmission, another distinctive feature of the Stercoraria. Once in the vertebrate host, the trypanosomes continue their life cycle, usually extracellularly (though intracellularly in one important species), by undergoing binary fission. It is a third characteristic of the Stercoraria that multiplication in the vertebrate host occurs only at certain phases of the life cycle, and that the individuals which divide are not in the trypomastigote form. Trypomastigote individuals develop from the products of these periodic bouts of multiplication.

There are three subgenera in this section, *Megatrypanum*, *Herpetosoma* and *Schizotrypanum*.

Subgenus *Megatrypanum*

These are non-pathogenic species parasitizing ruminants; none can infect man. The trypomastigote forms in the vertebrate host are large ($50\ \mu$ or more in length), have a long pointed posterior end extending well behind the kinetoplast, and a free flagellum (i.e. the flagellum extends beyond the anterior end of the body). Species

---

[1] It is probable that trypanosomes can penetrate an intact membrane, though this is not certain. It may be that they depend on the presence of small lesions, as they do to pass through the skin.

include *T. (Megatrypanum) theileri* of cattle (Fig. 11), transmitted by horse-flies (Diptera, Tabanidae) and *T. (M.) melophagium* of sheep, transmitted by louse-flies (Diptera, Hippoboscidae), both of which occur throughout the world, including Britain. They are usually present in the blood of their vertebrate hosts in very small numbers, which makes them very difficult to find in blood films (see Chapter 13). *T. (M.) theileri* divides in the vertebrate as epimastigote forms; division of *T. (M.) melophagium* in the vertebrate host has not been seen.

Subgenus *Herpetosoma*

Most species of this subgenus are non-pathogenic parasites of rodents. The trypomastigote forms in the blood of the vertebrate host are similar in all these species, being about 30 μ long, with a pointed posterior end extending some distance behind the kinetoplast, and a free flagellum. The nucleus is often noticeably in front of the mid-point of the body. Most species are very host-specific, being unable to infect even closely related vertebrate species: e.g. *T. lewisi* of the rat cannot infect mice, and *T. musculi* of mice cannot infect rats, although morphologically the two parasites are indistinguishable

Important species. (1) *T. (Herpetosoma) rangeli* (Fig. 12). This species infects man as well as dogs, cats, opossums and monkeys in parts of South America. Though morphologically resembling others in this section, it is exceptional in that its development in its vector, the bug *Rhodnius prolixus* (Hemiptera, Reduviidae) takes place, after an initial phase in the mid-gut, in the haemocoel, and metacyclic trypanosomes develop in the bug's salivary glands, whence they are inoculated into the next vertebrate on which the bug feeds (this type of inoculative transmission is characteristic of the second section of the trypanosomes of mammals, the Salivaria). Sometimes transmission occurs via the hind-gut, by contaminative methods, also. It is not certainly known whether *T. rangeli* divides at all in its vertebrate host, or, if it does, where and in what form division occurs. *T. rangeli* seems to be non-pathogenic to its vertebrate hosts, though there is some evidence that it is harmful to its invertebrate host (Levine, 1961), apparently unique behaviour for a trypanosome.

(2) *T. (H.) lewisi* (Fig. 13). This is a cosmopolitan non-pathogenic parasite of the wild rat (*Rattus norvegicus* and *R. rattus*). Its vectors are the rat fleas *Xenopsylla cheopis* and *Ceratophyllus fasciatus* (Siphonaptera), in which it develops in the mid- and hind-gut.

Reports are conflicting, but in some circumstances, at least, it has an intracellular phase of development in a mid-gut cell. After introduction (by contaminative methods) of the metacyclic forms into the rat, division (multiple fission of epimastigote forms) occurs in the blood capillaries of the viscera, especially the kidneys, for 4–5 days before flagellates are seen in the blood. At first these are of various shapes and sizes, but about 7–10 days after its infection the rat produces an antibody ('ablastin'), which inhibits further division and also probably selectively kills the dividing forms. Thereafter only 'adult' trypomastigote stages are seen in the blood, until eventually (usually one month later) the rat secretes a second antibody which kills all the trypanosomes (Ormerod, 1963). Such a cured rat can, sometimes at least, be reinfected, but after two or three re-infections a complete immunity develops which apparently lasts for the remainder of the animal's life. (This is what happens in the laboratory rat. The infection in wild rats has been less carefully studied.) This trypanosome is often maintained in laboratories for teaching or research purposes. It cannot infect man.

Other species include *T. (H.) musculi* of mice, *T. (H.) nabiasi* of rabbits, *T. (H.) microti* of voles (all cosmopolitan, including Britain) and *T. (H.) primatum* of the African chimpanzee.

Subgenus *Schizotrypanum*

The only really well-defined species of this subgenus is *T. (S.) cruzi*, an important parasite of man in South and Central America. There are many reports of similar organisms from various mammals (opossums, armadillos, rodents, monkeys) in South and Central America and the southern USA; whether all, any or none of these are identical with *T. cruzi* is uncertain at present.

T. (Schizotrypanum) cruzi infects man all too commonly throughout South America (especially Brazil and Argentina); in Central America and Mexico it is less common, and in the southern USA only two human infections have been reported (Bray, in Weinman and Ristic, 1968, p. 330). In all these areas, however, and as far north as Maryland, what is almost certainly the same parasite has been recorded from various wild animals, chiefly rodents and racoons in USA, vampire bats in Panama, and armadillos, opossums and wild guinea-pigs in South America. Dogs and cats are often important sources of human infections. The vectors are bugs, *Rhodnius prolixus* and others (Hemiptera, Reduviidae), which become infected by feeding on vertebrates (human or other) with parasites in their blood. The human disease (Chagas's disease) is really one of poverty—

being associated with squalid living conditions, in bug-infested mud-walled huts. Development in the vector is typical of the Stercoraria, the metacyclic trypanosomes being passed out in the bug's faeces and entering the vertebrate by one or other of the contaminative routes outlined above. The bugs feed at night, and sleepy children, disturbed by the insects, often scratch the region of the bite, thus collecting infective faeces on their fingers, and then rub their eyes, so introducing metacyclic trypanosomes via their conjunctiva. In young children the disease may be rapidly fatal, but more often it is chronic; the parasites multiply in the vertebrate as large amastigote forms (about $4\,\mu$ in diameter) in various lymphoid–macrophage cells and also muscle cells—including those of the heart (Fig. 14). Thus chronic sufferers from Chagas's disease may die from heart failure. The amastigote forms then elongate, develop a flagellum, and metamorphose, via promastigote and epimastigote stages, into trypomastigote forms (Figs. 15–17). These enter the blood, in which they circulate for a while. They are rather small (about $20\,\mu$ long), with an acutely pointed posterior end, very large kinetoplast, and free flagellum (Fig. 18). On dried blood films they often lie curved in the shape of a letter C. After a time, they re-enter suitable cells, become amastigote and commence dividing once more. No effective drug is in use for curing or preventing Chagas's disease at present, though some are being tested.

Section B. Salivaria

Almost all the species forming this group are transmitted by tsetse flies (Diptera, Glossinidae);[1] the few exceptions have evolved from forms which had a tsetse vector. They form a fairly closely knit group, differing in several respects from members of the Stercoraria. In the vector, these species undergo an initial period of multiplication in the trypomastigote form, before changing into the epimastigote form; the developmental cycle ends with the production of metacyclic trypomastigote forms in the fore-gut (salivary glands or proboscis) of the tsetse fly, where they are well placed to be inoculated into the blood of the next host on which the fly feeds. A fly, once infected, remains so for the remainder of its life. The trypomastigote forms in the blood of the vertebrate host have a less elongated posterior end than do the Salivaria; multiplication is by binary

---

[1] Tsetse flies (*Glossina* spp.) are large Diptera, both sexes of which feed only on blood. The female deposits larvae singly about every ten days during her life, and the larvae pupate immediately. Tsetse flies are found only in tropical Africa. For further details, see Buxton, 1955.

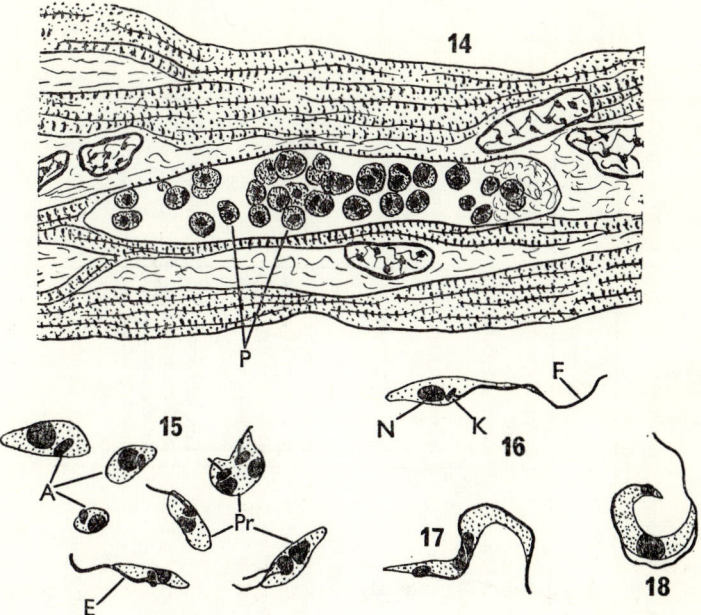

Fig. 14. *T. cruzi*; amastigote forms in a pseudocyst seen in a section of heart muscle from a mouse. ×1,500.

15. *T. cruzi*; amastigote, promastigote and epimastigote forms seen in a smear made from heart muscle of an infected mouse. ×1,500.

16. *T. cruzi*; epimastigote form from a culture. ×1,500.

17. *T. cruzi*; metacyclic trypomastigote form from an infected reduviid bug. ×1,500.

18. *T. cruzi*; trypomastigote form from blood of a mouse. ×1,500.

Abbreviations: A – amastigote; E – epimastigote; F – flagellum; K – kinetoplast; N – nucleus; P – parasite; Pr – promastigote.

fission of the trypomastigotes in the circulating blood—there are no intracellular stages, nor are dividing forms restricted to visceral capillaries as they are in *T. lewisi*. Most species of this section are restricted to Africa, where the group as a whole must have evolved; a few species have spread beyond this continent, and are no longer transmitted by tsetse flies. Most members of this section are important pathogens, either to man or to his domestic animals. Several sub-genera are recognized by Hoare (1966). The geographical distribution and hosts of the various species are shown in Table 3.

TABLE 3

GEOGRAPHICAL DISTRIBUTION, VERTEBRATE HOSTS AND TRANSMISSION OF SALIVARIAN TRYPANOSOMES

Species	Geographical distribution	Main vertebrate hosts Wild	Main vertebrate hosts Domestic	Laboratory	Transmission Cyclical	Transmission Non-cyclical
T. v. vivax	W., Central and E. Africa	Waterbuck, reedbuck, eland, giraffe, bushpig	Cattle, sheep, goats, horses, donkeys (often pathogenic)	Normally none (rats with serum supplement)	*Glossina* spp.	Biting Diptera
T. v. viennei	Mauritius, W. Indies, S. America	?	As for *T. v. vivax*	Normally none	None	Biting Diptera
T. uniforme	W. Uganda, E. Congo (Tanzania and Zululand rarely)	Antelope	Cattle, pigs (rarely if ever pathogenic)	Normally none	*Glossina* spp.	Biting Diptera
T. congolense	W., Central and E. Africa	Many antelope, giraffe, eland, wildebeest	Cattle, sheep, horses, donkeys, pigs, dogs, rarely camels (often pathogenic)	Rodents (not all strains are infective; pathogenicity varies)	*Glossina* spp.	Biting Diptera (less important)
T. simiae	W., Central, E. and parts of S. Africa	Warthog	Pigs (very pathogenic)	*Cercopithecus* monkeys (very pathogenic) and splenectomized rabbits	*Glossina* spp.	Biting Diptera (less important)
T. suis	Tanzania, Rwanda-Burundi, ? Congo	Warthog	Pigs (pathogenic to young)	None known	*Glossina* spp.	?

Species	Geographical distribution	Main vertebrate hosts Wild	Main vertebrate hosts Domestic	Laboratory	Transmission Cyclical	Transmission Non-cyclical
T. b. brucei	W., Central, E. and parts of S. Africa	Many antelope, warthog, wildebeest	Camels, horses, donkeys, dogs (fatal), cattle, pigs, sheep, goats (may be pathogenic but more chronic)	Rodents, rabbits, monkeys (pathogenic)	*Glossina* spp.	Biting Diptera (unimportant)
T. b. rhodesiense	Central and E. Africa	Bushbuck, ?other antelope	Cattle, ?others, MAN (pathogenic)	As for *T. b. brucei*	*Glossina* spp.	Biting Diptera (unimportant)
T. b. gambiense	W. and Central Africa	Probably none	?Pig, MAN (pathogenic but more chronic than *T. b. rhodesiense*)	As for *T. b. brucei* (less pathogenic)	*Glossina* spp.	Biting Diptera (unimportant)
T. e. evansi	N. Africa, Asia, S. China, Philippines, Mauritius, Central and S. America (eliminated from N. America and Australia)	Tapir (in America), deer (in Mauritius), vampire bat (S. America)	Camels, horses, donkeys, mules (often pathogenic); cattle, water buffalo, sheep, goats, dogs, Indian elephant	Rodents (pathogenic)	None	Biting Diptera
T. e. equiperdum	S. America mainly (previously also Europe, India, N. America)	None	Equines (pathogenic)	Rabbits (intra-testicular or intrascrotal), dogs (some strains)	None	Coitus
T. equinum	S. America (and ?Sudan)	Vampire bat and capibara (in S. America)	Horses (pathogenic); donkeys, mules (less pathogenic); cattle, sheep, goats, pigs (chronic)	Rodents (pathogenic)	None	Biting Diptera

Subgenus *Duttonella*

The trypomastigote forms in the vertebrate host typically have a rather broad, rounded posterior end, giving them a 'club-shaped' appearance, with a large round kinetoplast which is usually terminal in position; a free flagellum is present. In the tsetse fly, development (which follows the course outlined above) occurs exclusively in the insect's proboscis. In the laboratory, fairly high infection rates can be obtained in insects fed on infected vertebrates—up to about 50 or 60%. There are two species, differentiated by their size.

(1) *T. (Duttonella) vivax*. The forms in the blood of the vertebrate host measure from 20 to 26 μ in length, with an average of more than 20 μ (Fig. 19). This is a common and pathogenic parasite of domestic cattle and sheep throughout Africa, between the southern limit of the Sahara and the tropic of Capricorn. It is also found in a variety of wild animals. *T. (D.) vivax* has also spread to Mauritius, the West Indies and parts of South America, presumably being carried there in domestic animals. In these regions it is transmitted by biting flies (Diptera), chiefly of the genera *Tabanus* and *Stomoxys*, in which it does not undergo any cyclical development, being merely carried passively in the proboscis of the insect. This type of transmission is called non-cyclical transmission (see p. 13); the insect functions only as a flying hypodermic syringe, and can remain infective only for as long as the blood in its proboscis remains moist—quite a short time. Thus non-cyclical transmission depends upon a fly's being disturbed while in the act of feeding and flying to another nearby animal to complete the feed. Hoare (1967) has suggested that this form of *T. vivax* should be regarded as a separate subspecies—*T. vivax viennei*.

(2) *T. (D.) uniforme*. Morphologically very similar to *T. vivax*, this species is smaller; the forms in the blood of the vertebrate measure only 12 to 24 μ, with a mean length less than 18 μ (Fig. 20). It is more restricted in its distribution than *T. vivax*, having been reported only from a few parts of tropical Africa. Its hosts are similar to those of *T. vivax*, but it seems to be rarely, if ever, pathogenic.

Subgenus *Nannomonas*

As seen in the vertebrate host, members of this subgenus are rather small trypomastigote forms, lacking a free flagellum (i.e. the flagellum does not extend beyond the anterior tip of the cytoplasm); however, on dried blood-films there may appear to be a short one. The kineto-

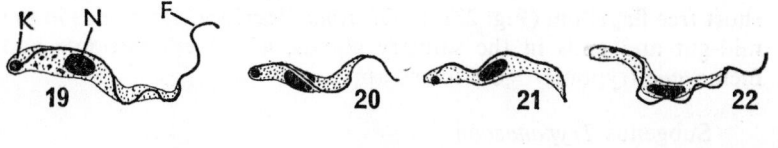

Fig. 19. *T. vivax*; trypomastigote form from blood of a Nigerian sheep. ×1,500. (From a slide kindly supplied by Dr D. G. Godfrey.)
  20. *T. uniforme* in blood of a calf from the Congo. ×1,500. (From a slide made by Dr L. van Hoof in 1938 and kindly supplied by Dr C. A. Hoare.)
  21. *T. congolense* in blood of a laboratory rodent. ×1,500.
  22. *T. simiae* in blood of a pig. ×1,500.
  23. *T. suis* in blood of a pig. ×1,500.
Abbreviations: F – flagellum; K – kinetoplast; N – nucleus.

plast is small and usually marginal. In the tsetse fly, development begins in the mid-gut, and is completed in the proboscis, where the epimastigote and metacyclic trypomastigote forms are found. Infection rates in *Glossina* are not very high—even in the laboratory usually only 20–30%. The subgenus is restricted to tropical Africa.

(1) *T. (Nannomonas) congolense* (Fig. 21). The length of the forms seen in the blood of the vertebrate host differs considerably in different strains; some workers consider the longer strains to constitute a distinct species, *T. (N.) dimorphon*. The wild vertebrate hosts of *T. congolense* are mainly antelope. It is often an important pathogen of domestic animals.

(2) *T. (N.) simiae*. This species is difficult to distinguish morphologically from *T. congolense*. Generally some of the forms in the vertebrate are longer, with a better developed undulating membrane. Other individuals are indistinguishable from *T. congolense* (Fig. 22). This parasite seems to occur only in pigs, wild and domestic; to the latter at least it is extremely pathogenic. It was named *T. simiae* because it is also very pathogenic to experimentally infected monkeys.

Subgenus *Pycnomonas*

This subgenus includes only one species, *T. (Pycnomonas) suis*, a rare parasite of central African wild and domestic pigs. The forms in the vertebrate are short and broad with a pointed posterior end and a short free flagellum (Fig. 23). In *Glossina,* development begins in the mid-gut and ends in the salivary glands, where epimastigote and metacyclic trypomastigote forms appear.

Subgenus *Trypanozoon*

This subgenus seems to be in the midst of a period of rapid evolution at the present time, which makes it difficult for systematists to divide it up neatly into distinct species. The 'classical' view is that there are six species in the subgenus; some of these, however, are probably better regarded as subspecies. There are two clearly separated groups within the subgenus—those which are transmitted cyclically by tsetse flies (some non-cyclical transmission by biting Diptera doubtless occurs in this group, but it is of minor overall importance) and those which are not. The former group consists 'classically' of three species, *T. brucei, T. rhodesiense* and *T. gambiense:* however, the latter two are probably best regarded as subspecies of the first (it has been argued that they should not be distinguished to even this extent (Ormerod, 1967)). In the second group are *T. evansi, T. equiperdum* and *T. equinum,* of which *T. equiperdum* is probably best regarded as a subspecies of *T. evansi* (see Hoare, 1967).

T. (T.) brucei (including its subspecies) exists in the blood of its mammalian hosts (Table 3) in two morphological forms[1]—long, slender trypomastigote individuals (about $30 \times 1 \cdot 5 \mu$) with a distinct free flagellum, and short, stumpy trypomastigotes (about $18 \times 3 \cdot 5 \mu$) with no free flagellum (Figs. 24–8). These two forms are inter-convertible in at least one direction (slender→stumpy), and a complete range of intermediates is seen. The proportions of the two forms seen vary widely. *T. brucei* usually gives rise to relapsing infections in man and laboratory rodents, in which peaks of parasitaemia are followed by the production of antibody by the host which kills most (but not all) of the trypanosomes. The survivors of this crisis differ antigenically from the bulk of the population before the crisis; thus they are unaffected by the antibody and increase in numbers again until the host has produced a second lot of antibody, specific to this antigenic type; and so on. Ultimately the host always loses this battle. There appears to be virtually no limit to the number of antigenic variants the trypanosomes can produce, though the mechanism underlying this is not understood. During the periods of increasing parasitaemia, leading up to a crisis, trypanosomes are predominantly of the long, slender type, this being the only type

[1] A phenomenon known as *pleomorphism* (or, less correctly, polymorphism).

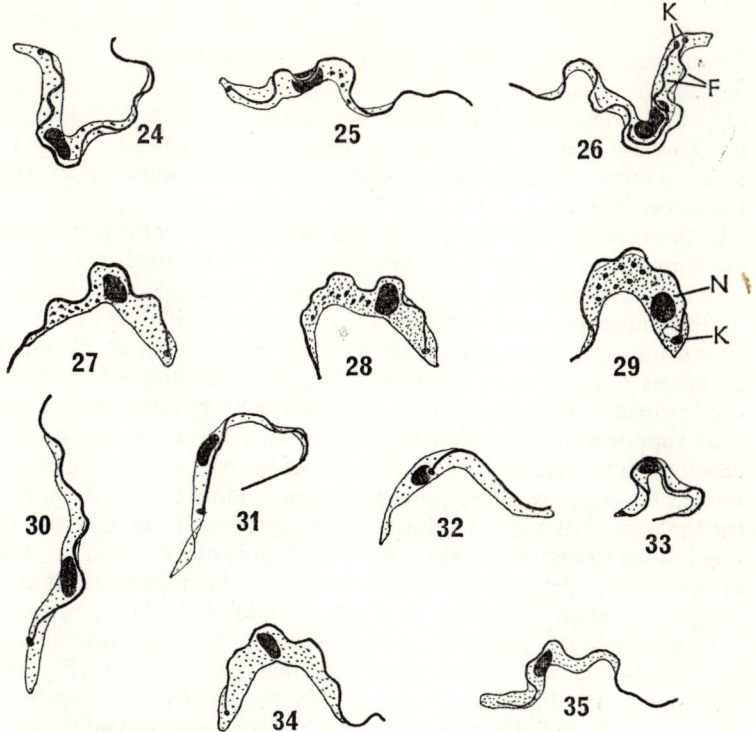

Figs. 24–33. *T. brucei* sspp. All ×1,500.
    24 and 25. Slender forms from blood of rat.
    26. Dividing slender trypomastigote form from blood of rat.
    27. Intermediate form from blood of rat.
    28. Stumpy trypomastigote form from blood of rat.
    29. Posteronuclear stumpy form from blood of rat.
    30. Elongate trypomastigote form from mid-gut of *Glossina morsitans*.
    31. Elongate 'ribbon-like' trypomastigote form from proventriculus of *G. morsitans*.
    32. Epimastigote form from salivary gland of *G. morsitans*.
    33. Metacyclic trypomastigote form from salivary gland of *G. morsitans*.
    34. *T. evansi evansi* from blood of a Sudanese camel. ×1,500.
    35. *T. e. equinum* (='akinetoplastic strain of *T. e. evansi*') from blood of rat. ×1,500.

Abbreviations: F – flagellum; K – kinetoplast; N – nucleus.

which divides; as the crisis approaches, stumpy forms become more numerous, and during and immediately after it they may predominate. Sometimes a form of stumpy individual develops, in which the nucleus lies in the posterior third of the body, close to the kinetoplast (Fig. 29). The significance of these 'posteronuclear' (or PN) forms is not known. There is some, though not conclusive, evidence that the stumpy forms are those which continue the developmental cycle in the tsetse fly.

It has recently been found that the slender forms of this subgenus are unusual in having an aerobic system of oxidative metabolism which does not involve the Krebs's cycle-cytochrome chain enzymes of the mitochondrion (see pp. 36–7 above); thus, in them, the mitochondrion is non-functional. The stumpy forms, however, as well as all stages developing in the vector with the exception of the metacyclic forms, undergo normal mitochondrial terminal respiration (this supports the view that it is the stumpy forms which continue development in the vector). In the slender forms, glucose is degraded only as far as pyruvic acid, and consequently more oxygen is required for the release of a given amount of energy than in the conventional breakdown to carbon dioxide and water; thus, presumably, this type of respiration depends on the presence of a high concentration of oxygen in the environment, as obtains in blood (Vickerman, 1965).

In the tsetse fly, *T. brucei* undergoes a complicated developmental cycle, in which very elongated trypomastigote forms (Fig. 30) multiply in the insect's mid-gut within the peritrophic membrane (a chitinous tube lining the mid-gut); they then pass out of the open hind end of the peritrophic membrane, at the junction of the mid- and hind-gut, and migrate forwards between the membrane and the gut wall to the proventriculus (the anterior part of the mid-gut), where the peritrophic membrane is secreted. They pass through the membrane while it is still soft (Fig. 31), enter the lumen of the proventriculus, and then travel (it is thought) down the proboscis and back up the salivary duct to the salivary glands. In the glands, epimastigote forms (Fig. 32) develop and multiply and subsequently metamorphose into metacyclic trypomastigote forms (Fig. 33) which are injected into the next vertebrate on which the fly feeds, and in which (if it is a susceptible species of mammal) they change into the slender blood forms and begin an infection (Buxton, 1955). Only a small proportion of tsetse flies which ingest *T. brucei* develop salivary gland infections, which is perhaps not surprising in view of the complicated migrations undergone by the trypanosomes in the fly.

Two sub-species of *T. brucei*, *T. b. gambiense* and *T. b. rhodesiense*, infect man and cause African human trypanosomiasis ('sleeping sickness') throughout tropical Africa. *T. b. gambiense* occurs in west and central Africa, *T. b. rhodesiense* in the eastern half of that continent. Around the northern and eastern shores of Lake Victoria, the two subspecies meet and overlap. The disease produced in man by the two sub-species is basically similar, differing in that infections due to *T. b. gambiense* are much more chronic, lasting a matter of years in untreated patients, while those due to *T. b. rhodesiense* are measured in months. Both forms are invariably fatal, unless treated. Initially the trypanosomes live and multiply in the blood and tissue fluid, producing a febrile condition which may be quite mild (especially with *T. b. gambiense*). After a few months *(T. b. rhodesiense)* or a year or so *(T. b. gambiense)* the parasites invade the central nervous system and multiply in the cerebrospinal fluid; here, they ultimately cause brain damage which leads to the coma from which the disease gets the name 'sleeping sickness'. Death ensues.

In the early stage of the infection, before invasion of the central nervous system, the disease is readily curable (the most commonly used drug is suramin). In the late stage, treatment is less successful, but in the last decade or two the introduction of organic arsenical compounds based on melarsen (e.g. melarsen oxide/BAL or 'mel B') has greatly improved the outlook.

T. b. gambiense is usually transmitted to man by *Glossina palpalis* and related species, flies which live mainly along the banks of rivers; *T. b. rhodesiense* is usually transmitted by flies of the *Glossina morsitans* group, which inhabit vast areas of the drier plains in east Africa. This results in interesting differences in the epidemiology of the two diseases, which have been discussed by Ashcroft (1959). Experimentally, however, all species of *Glossina* can transmit either subspecies of *T. brucei* (and *T. b. brucei* itself).

The species *T. (T.) evansi* (including two subspecies, *evansi* and *equiperdum*) and *T. (T.) equinum* (see Tables 1 and 3) are thought to have evolved relatively recently from *T. b. brucei* by the carriage of the latter in infected animals (possibly camels) northwards in Africa beyond the range of *Glossina*. Here, the parasites became adapted to non-cyclical transmission by biting Diptera (cf. *T. vivax* in Mauritius, p. 56 above) and so survived and underwent speciation on their own, as follows (Hoare, 1956).

In strains of *T. brucei* which are transmitted by hypodermic syringes in laboratories, there is no longer any need for the ability to infect tsetse flies. Selection pressure operates against the retention of pleomorphism (see footnote on p. 58) in such strains, as in these circumstances the stumpy forms are not only unnecessary but, since they do not divide, actually reduce the strain's reproductive rate and thus diminish its chances of being transmitted to another vertebrate host and, hence, surviving. Therefore, such non-cyclically transmitted strains in laboratories quite rapidly (within a few years at the most) become monomorphic, all individuals being more or less of the slender type (Fig. 34). Such strains also lose the ability to infect tsetse flies, presumably because they have lost the ability to synthesize mitochondrial enzymes. This is presumably how *T. evansi* evolved, since it is usually monomorphic (being morphologically indistinguishable from a strain of *T. brucei* which has been maintained by syringe-passage in a laboratory for some years) and unable to infect tsetse flies: in this instance the biting Diptera of North Africa played the role of the laboratory syringe.

This hypothetical but plausible evolutionary story can be carried further. The kinetoplast, which, it will be remembered, is a mass of DNA within the mitochondrion of trypanosomatids, is presumably concerned, at least in part, with the synthesis of the mitochondrial enzymes. Thus in the slender forms of *T. brucei* it is redundant, at least as far as this function is concerned; and it is common to find a small proportion of the slender forms of this species (about 1%) lacking a kinetoplast, these being presumably mutants. In strains being transmitted cyclically by tsetse flies, such akinetoplastic individuals would be 'filtered out' by their inability to synthesize the necessary respiratory enzymes to enable them to develop in the insect. However, in *T. evansi* this particular selection pressure would be lacking, and indeed strains of this species are often found with quite high proportions of akinetoplastic individuals. Sometimes, particularly in South America, strains are found in which all the organisms lack a kinetoplast (Fig. 35). These are generally referred to as a distinct species, *T. equinum*.

T. evansi equiperdum[1] is morphologically identical with *T. evansi* but has become specialized in another direction; it is the only known member of the genus which is not dependent upon an invertebrate vector to transmit it from vertebrate to vertebrate. It is a parasite of horses and donkeys, and is more numerous in oedematous patches of skin than in the blood; when such patches occur on the genitalia, they often become eroded during coitus and it is in this way that the trypanosomes are transmitted.

[1] Although usually treated as a distinct species, the differences between this parasite and *T. evansi* are only behavioural, and so it should probably be regarded as a subspecies of *T. evansi* (see Hoare, 1967).

Diagnosis of trypanosome infections

The simplest method of diagnosing trypanosome infections of vertebrates is by making and examining thin films of peripheral blood, stained with Giemsa's stain (Chapter 13). Alternatively, a drop of blood can be mounted on a slide beneath a coverslip and examined immediately for motile trypanosomes, at a magnification of about × 400. If the parasites are scanty, they may not be found by these methods. Some concentration of parasites in the blood of mammals can be achieved by making stained thick blood films (Chapter 13). The most sensitive method of diagnosis, for most species of *Trypanosoma,* is to collect some blood aseptically and inoculate it to blood-agar cultures (Chapter 13); trypanosomes, if present, will multiply as (usually) epimastigotes. However, the Salivaria are difficult to cultivate in this way and special media must be used (Taylor and Baker, 1968).

Apart from these methods, human infections with *T. cruzi* may be detected by feeding reduviid bugs, which have been bred in the laboratory, on a suspect and then seeing whether the bugs become infected. This method is known as *xenodiagnosis*. *T. b. gambiense*, which may be very sparse in the peripheral blood of infected persons, is often more readily found in the 'juice' obtained by puncturing lymph glands. In the later stages of infection with both *T. b. gambiense* and *T. b. rhodesiense,* the organisms may be found more readily in cerebrospinal fluid, obtained by lumbar puncture, than in blood. Inoculation of blood (or other material) to susceptible laboratory animals such as rats or mice is also used (though *T. b. gambiense* is only slightly, if at all, infective to these animals).

4

PARASITIC FLAGELLATES OF THE ALIMENTARY AND URINOGENITAL TRACTS

Six of the nine orders of the class Zoomastigophorea (superclass Mastigophora, subphylum Sarcomastigophora; see pp. 17–18) are mostly or entirely parasitic and one other order contains a few parasitic forms. One of these orders, the Kinetoplastida, forms the subject of the preceding chapter; the rest will be considered briefly here, together with the enigmatic organisms classified by the Society of Protozoologists' Committee (Honigberg *et al.*, 1964) as the superclass Opalinata (see p. 75 below).

CLASS Zoomastigophorea
Many parasitic genera are grouped within the seven orders of Zoomastigophorea referred to above. Very few of them are of economic importance, and probably only four genera *(Histomonas, Hexamita, Giardia, Trichomonas)* contain pathogenic species; some, in the order Hypermastigida are not only useful but essential to their hosts (termites and certain Orthoptera or wood-roaches). Most species are transmitted directly, by the voiding of parasites (often encysted) in the faeces and their ingestion by a new host; exceptions to this simple system will be mentioned below. Sexual reproduction is known in only a few genera of the orders Oxymonadida and Hypermastigida.

Order Rhizomastigida
This order contains the 'amoeboflagellates', organisms which at the same or different stages of their life history possess both pseudopodia

and flagella.[1] A few species of a few genera are parasitic in amphibian and insect larvae, and one in birds: *Histomonas meleagridis*, which is the only one of any importance.

Genus *Histomonas* Tyzzer, 1920

Only one species is known in this genus—*H. meleagridis*. It inhabits the intestinal caeca (both lumen and mucosa) and the liver of gallinaceous birds, including chickens and turkeys, and has been reported from all parts of the world. In chickens it is seldom pathogenic, and extremely common. In turkeys it is one of the more important pathogens, producing a disease called histomoniasis or 'blackhead' (because the head of an infected bird may become darkened). Recent electron microscope study of the structure of *H. meleagridis* suggests a relationship to the trichomonads (see below) (Schuster, 1968).

Morphology and life cycle

H. meleagridis (Fig. 36) is an amoeboid organism which may or may not also be flagellate. The amoeboid stage is found only in the host's tissues, and is usually oval, varying from $8-21 \times 6-15$ μ in size. The flagellate phase occurs at times in the lumen of the caeca (and in cultures *in vitro*); it too is amoeboid, from 5–30 μ in diameter, and possesses usually one (sometimes as many as four) short flagella, arising from a basal body (or bodies) close to the single, eccentric nucleus. The parasites divide by binary fission, no sexual process being known.

The trophozoites do not survive well outside the body, and no cyst is produced. Transmission occurs by means of trophozoites carried inside the eggs of the parasitic nematode *Heterakis gallinae*. By no means all of the worm eggs are infected—possibly only about 0·1 % (Kendall, 1959). However, infected eggs can survive in soil for several years (Levine, 1961).

Epidemiology

As stated above, *Histomonas* is very common in chickens, to which it is rarely pathogenic. *Heterakis* is also common in these birds. It is probable that turkeys usually become infected by ingesting infected eggs deposited by chickens. Wild gallinaceous birds (e.g.

[1] It is to this order that *Dientamoeba* (see p. 76) would belong if the suggestion that it is an aberrant flagellate were true.

pheasant, quail, grouse) may also serve as reservoirs of infection sometimes. (For further details, see Reid, 1967.)

Pathogenesis

When the parasites (in the amoeboid stage) invade and multiply in the caecal mucosa, they produce much tissue damage resulting in ulcers which may become widespread. The infected caeca become inflamed, enlarged and filled with a yellowish, hard exudate; the lesions may perforate and cause peritonitis. A yellow diarrhoea develops. Trophozoites are often carried by the blood-stream to the turkey's liver (and, less commonly, to other organs), where they produce circular, yellowish-green abscesses (areas of necrosis) up to 1 cm or more in diameter. Trophozoites are readily found in sections of both caecal and hepatic lesions.

If untreated, death may occur rapidly in 50–100% of young turkeys, less in older birds. Birds which do recover are immune to re-infection (Levine, 1961).

Diagnosis

Apart from the clinical picture, histomoniasis can be diagnosed from the appearance of the hepatic lesions at autopsy, and confirmed by demonstrating the organisms in sections of, or scrapings from, caecal or hepatic lesions.

Treatment and prevention

Several drugs, usually given mixed with the food, may be used to suppress the infection; as radical cure is seldom achieved treatment may be necessary throughout the birds' life. Reasonably effective drugs include 4-nitrophenylarsonic acid, 2-amino-5-nitrothiazole and derivatives and, as a prophylactic only, furazolidone. Turkeys should not be reared together with chickens, nor on land previously used by chickens, to reduce the likelihood of their ingesting infected eggs of *Heterakis*. For more details, see Richardson and Kendall (1963).

Order Retortamonadida

Several genera of small parasitic flagellates with two, four or more flagella are included in this order. They have a cytostome, are not bilaterally symmetrical and lack an axostyle and an undulating

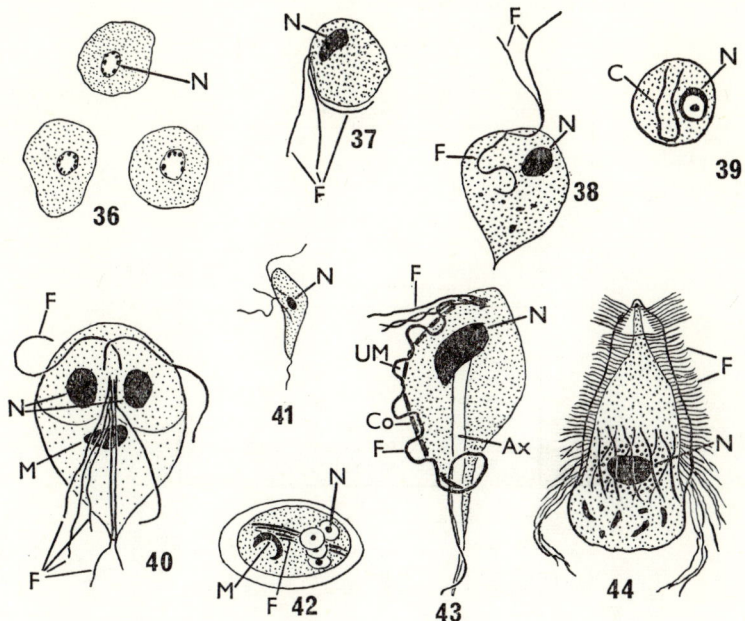

Fig. 36. *Histomonas meleagridis;* amoeboid forms from a section of turkey liver. ×1,500. (From a slide kindly supplied by Dr S. B. Kendall.)
    37. *Retortamonas hominis;* trophozoite from a culture. ×1,500.
    38. *Chilomastix mesnili;* trophozoite from human faeces. ×1,500.
    39. *C. mesnili;* cyst from human faeces. ×1,500.
    40. *Giardia lamblia;* trophozoite (plan view) in smear of human faeces. ×1,500.
    41. *Giardia* sp.; trophozoite (side view) in section of rabbit intestine. ×1,500.
    42. *G. lamblia;* cyst in smear of human faeces. × 1,500.
    43. *Trichomonas muris* in smear of rodent faeces. ×1,500.
    44. *Trichonympha campanula* from gut of termite. × 150. (Redrawn and slightly modified from Mackinnon and Hawes, 1961, fig. 83.)

Abbreviations: Ax – axostyle; C – cytostomal filament; Co – costa; F – flagellum; M – median body; N – nucleus; UM – undulating membrane.

membrane. None is of economic importance, nor, as far as is known, pathogenic. Their hosts may be vertebrate or invertebrate. Further details of those which infect man and domestic animals are given by Levine, 1961. Two species inhabit the intestine of man, *Retortamonas intestinalis* and *Chilomastix mesnili* (see Table 4 and Figs. 37–9).

TABLE 4
PARASITIC FLAGELLATES INHABITING THE HUMAN INTESTINE

Species	Trophozoite			Cyst		
	Size (μ)	No. of flagella	Other characters	Size (μ)	No. of nuclei	Other characters
Chilomastix mesnili	6–24 × 3–10	3 + 1[1]	Cytostome	7–10	1	Lemon-shaped; cytostome fibril
Enteromonas hominis[2]	4–10 × 3–6	3 + 1	No cytostome	6–8 × 4–5	4	None
Retortamonas intestinalis	4–9 × 3–4	1 + 1	Cytostome	4·5–6	1	As *C. mesnili*
Giardia lamblia[3]	10–20 × 5–10	4 pairs	Sucker; 2 nuclei; median bodies	8–14 × 6–10	4	Fibrils and axonemes
Trichomonas hominis	8–20 × 3–14	5 + 1	Undulating membrane; axostyle; cytostome	No cyst		

[1] This notation implies that there are three free anterior flagella and one recurrent one.
[2] The taxonomic position of this organism within the Zoomastigophorea is doubtful.
[3] Pathogenic sometimes; described more fully in text (pp. 69–71).

Order Oxymonadida
These flagellates, which have four or more flagella and one to many nuclei and axostyles, are all symbiotic in the intestine of insects (Orthoptera, Blattidae) or termites (Isoptera) (see p. 79 below).

Order Diplomonadida
This order includes flagellates with a bilaterally symmetrical body, two nuclei and eight flagella. Almost all are parasitic in vertebrates or invertebrates. The two main genera, *Hexamita* and *Giardia,* are most easily distinguished by the fact that the former is narrower, lacks an anterior sucker, and has two pairs of flagella emerging from the body at the extreme front end; it does not encyst. *Giardia* is described below. *Hexamita* spp. inhabit the intestines of amphibia, fish, birds, mammals (not man) and orthopterous insects (Kudo, 1966). *H. salmonis* (of fish), *H. meleagridis* (of turkeys) and *H. columbae* (of pigeons) are pathogenic (for further details, see Levine, 1961 and Richardson and Kendall, 1963).

Genus *Giardia* Kenstler, 1882
Species of this genus have been described from the small intestine of man, dogs, cats, cattle, various rodents, rabbits and other mammals (see Levine, 1961); a few species are recorded from amphibia and reptiles. How many of these are really valid species is unknown, as is the pathogenicity of most of them. However, *G. canis* of the dog probably sometimes behaves as a pathogen, and *G. lamblia* of man certainly does so at times. The general morphology of all species is similar; for detailed morphological descriptions of *G. muris* (using light microscopy) see Filice (1952) and *G. intestinalis* (using electron microscopy) see Cheissin (1964, 1965). All species are transmitted by means of cysts passed out in the faeces.

G. lamblia. There is much confusion about this parasite's correct name (see Filice, 1952): it is also known as *G. intestinalis, G. duodenalis* and even as *Lamblia intestinalis.* It has the distinction of being probably the first parasitic protozoon to be seen—by Antony van Leeuwenhoeck in 1681, in his own faeces.

G. lamblia inhabits the lumen of the duodenum and upper ileum of man, monkeys and pigs in all parts of the world. In man it is common (about 10%), especially in children. It can cause a disease known as giardiasis or lambliasis.

Morphology and life cycle

The trophozoite is shaped like a pear bisected longitudinally (Figs. 40, 41). The flat (ventral) surface of the thick, broad (anterior) end forms a sucker, with a thickened anterior rim. With the aid of this the organism attaches itself to the intestinal mucosa. Above the sucker are two nuclei, and between the latter lie eight basal bodies giving rise to eight flagella. Only two of the flagella emerge directly from the body; two cross over and follow the front edge of the sucker before emerging laterally, and the remaining four run backwards in the body for some way—two emerging at the hind end of the sucker, and two at the extreme hind end of the body. By means of these flagella the trophozoite can swim actively. In the posterior half of the body lie one or (usually) two curved 'median bodies' of unknown function (often called parabasal bodies, but not morphologically the same as a true parabasal when examined by electron microscopy). The trophozoites usually measure from 10–20 μ long, 5–10 μ broad, and 2–4 μ thick. Reproduction is by binary fission; no sexual stage is known.

The cysts of *G. lamblia* are oval, about 8–14 μ long and 6–10 μ wide, and contain (when mature) four nuclei grouped at one end. The remains of the median body, flagella and the anterior rim of the sucker form a rather confused collection of fibrils within the cyst (Fig. 42), which makes it easy to recognize the cyst in faecal preparations. When a cyst is swallowed by a susceptible host, it presumably hatches in the duodenum. The quadrinucleate organism which emerges then divides into two binucleate trophozoites.

Pathogenesis

G. lamblia does not invade the tissues, but heavy infections may produce acute but not bloody diarrhoea, especially in children, and epigastric pain. The parasites are thought sometimes to swim up the bile duct and into the gall bladder where they may produce jaundice, nausea and vomiting. Heavy infections may also interfere with the absorption of fat (and fat-soluble vitamins) from the intestine.

Diagnosis

Diagnosis is confirmed by finding cysts, which may be very numerous, and—in cases with diarrhoea—trophozoites in faecal specimens (see pp. 152–155 below).

Parasitic flagellates of the alimentary and urinogenital tracts 71

Treatment and prevention

Giardiasis is readily cured by either of the antimalarial drugs mepacrine or chloroquine. Prevention is solely a matter of hygiene, particularly with regard to food, to prevent the ingestion of cysts (see *Entamoeba histolytica,* p. 83 below).

Order Trichomonadida

This order includes forms with, typically, three to five free anterior flagella and one recurrent flagellum which may be attached to the body to form an undulating membrane (see p. 30). Almost all are parasitic—in vertebrates or invertebrates (including termites, in which some species are truly symbiotic and help in digesting the wood on which the termite feeds but is itself unable to digest; see p. 74 below and also Kudo, 1966, pp. 457–74). As far as is known, no member of this order produces cysts, transmission being usually by the ingestion of trophozoites passed out in the faeces of infected hosts. The only economically important parasites in this order belong to the genus *Trichomonas*.

Genus *Trichomonas* Donné, 1837

Species of this genus occur in the intestines of mammals including man (Table 4), birds, reptiles, amphibia, molluscs (slugs) and termites; in the mouth of man and monkeys (*T. tenax*); and in the urinogenital tract of man and cattle (*T. vaginalis* and *T. foetus* respectively). No cyst is produced, and no sexual process is known. Reproduction is solely by binary fission. The following three species[1] are of economic importance.

(1) *T. vaginalis*. This organism seems to be exclusive to our own species (though hamsters can be infected experimentally). It lives in the female vagina and the male urethra or prostate, and is common throughout the world, particularly in women. Infection rates of up to 40% of unselected women have been recorded, and the incidence among those with vaginal upsets is higher—up to 70%. *T. vaginalis* may cause a mild disease (in women) called trichomonas vaginitis.

Morphology and life cycle

T. vaginalis is ovoid, narrower at the hind end, ranging in size from 10 to 30 μ long (usually averaging 14–17 μ) and 5 to 15 μ broad (Fig.

[1] Some authors divide the genus *Trichomonas* into three subgenera or even separate genera, on the basis of the number of anterior flagella. Species with three (including *T. foetus*) are called *Tritrichomonas*, those with four (including *T. vaginalis, T. tenax* and *T. gallinae*) are called *Trichomonas*, and those with five (including *T. hominis*) are called *Pentatrichomonas* (see Levine, 1961).

43). It has four free anterior flagella and one recurved posterior one which is attached to a thin, fin-like extension of the body to form an undulating membrane. The recurved flagellum and undulating membrane end about half-way along the body. All five flagella arise from basal bodies grouped at the anterior end, just in front of the single nucleus. With the aid of these flagella, the organism can swim actively. It also has a prominent skeletal axostyle of longitudinally arranged parallel microtubules (electron microscopy by Smith and Stewart, 1966), which may protrude from the hind end of the body; a cytostomal groove; a supporting costa at the base of the undulating membrane; and a true parabasal body (see p. 29) beside the nucleus. When seen alive, the parasite is easily identified by the characteristic and beautiful progression of waves backwards along the undulating membrane.

Transmission is probably via an infected male but may also occur by contamination of the vaginal orifice with infected material.

Pathogenesis

Most commonly the parasite is non-pathogenic; this is almost always so in males, though mild inflammation of the urethra may occasionally result. In women the organism is rather more commonly responsible for mild vaginal inflammation associated with a copious, foul-smelling discharge. It has been suggested that the parasite's pathogenicity is associated with endocrinal or other changes resulting in variation in the normal bacterial flora of the vagina, leading to a reduction in the acidity of its contents from the usual pH 4–4·5 to pH 5·5. However, the parasite cannot survive at neutrality (pH 7). *T. vaginalis* does not invade the tissues, as far as is known.

Diagnosis

Diagnosis is confirmed by isolating the parasites from vaginal discharge, best done by cultivation *in vitro* (Taylor and Baker, 1968). Microscopic examination is less reliable, though fresh preparations in which the organisms are motile are better than dried smears stained with Giemsa's (or other) stains (p. 160).

Treatment

Treatment is possible, though radical cure is sometimes not easily attained. Organic arsenicals such as carbasone (4-carbamylaminophenyl-arsonic acid) have been used, given both orally and in pessaries, but a newer non-arsenical compound (metronidazole), given orally, is probably at least as effective and less toxic. The antibiotic trichomycin is also effective, both orally and topically.

Further information about *T. vaginalis* can be obtained from a monograph by Trussell (1947).

(2) *Trichomonas foetus*. This species is a parasite of cattle and possibly other farm animals, living in the vagina and uterus of the female and beneath the prepuce of the male. In pregnant cows it may lead to abortion. It is widely distributed throughout the world and probably fairly common, but details of its distribution and incidence are not known.

Morphologically, *T. foetus* is basically similar to *T. vaginalis* but it has only three anterior flagella and the recurrent flagellum extends free for some distance beyond the hind end of the body. The parasite is transmitted during coitus (and sometimes by artificial insemination). If it is present in a pregnant uterus, an early abortion often results (usually between 1 and 16 weeks after fertilization), for the parasite invades the foetus. Diagnosis is confirmed in the same ways as in human *T. vaginalis* infection.

In cows the infection is usually self-limiting, all parasites being usually expelled with the aborted foetus and placenta. This is fortunate since no reliable treatment is known. Bulls remain infected for life, and hence are best destroyed or sold and not used for breeding. For further details and references see Levine (1961) and Richardson and Kendall (1963).

(3) *Trichomonas gallinae*. This species infects birds, primarily the domestic pigeon *Columba livia* but also other Columbiformes, turkeys, chickens and others. The parasites inhabit the mouth, pharynx, oesophagus, and crop. *T. gallinae* is common in pigeons but rare in chickens, and produces a disease variously known as 'canker', 'frounce', 'roup' or, more scientifically, avian trichomoniasis. The parasite resembles *T. vaginalis* but is smaller, $6-19 \times 2-9$ μ. There are four anterior flagella, and the recurrent flagellum ends about three-quarters of the way along the body.

T. gallinae produces a severe disease in young pigeons, which may be fatal. Almost all adult birds are infected but show no signs of disease. The organisms produce lesions in the anterior part of the digestive tract; the infection may spread to the sinuses of the head and to viscera such as the heart, lungs and liver. The lesions appear as small yellowish spots on the mucosa, and develop into large caseous nodules or masses. Young pigeons become infected from their parents during feeding (pigeons regurgitate food from their crop into the nestling's mouth).

Diagnosis is confirmed by finding parasites either on smears or in cultures made from lesions. The disease can be treated with 2-amino-5-nitrothiazole, one of the drugs used in treating histomoniasis (p. 66 above). The only way to keep young birds free from infection is to treat the adults before the eggs hatch.

Order Hypermastigida

These forms have a large number of flagella arising from the front end of the body, and live in the gut of insects—either termites (Isoptera) or roaches (Insecta, Orthoptera, family Blattidae). They are uninucleate but may have a very complex structure. Very large numbers may be present—sometimes up to one-third or one-half of the insect's weight may be accounted for by its flagellates (Kudo, 1966). Most of these flagellates can digest cellulose, and without them the wood-eating insects (termites and wood-roaches) would be unable to survive, since they themselves cannot digest the wood on which they feed. This is a truly symbiotic relationship, since the termite (or roach) obtains the food, the flagellate digests it, and both insect and protozoon live on it: neither can survive without the other. This also applies to some of the oxymonadid, diplomonadid and trichomonadid flagellates inhabiting the gut of termites, mentioned above (pp. 69 and 71).

In termites, one of the most important genera is *Trichonympha*—without it the insect cannot survive. Several other genera live in the termite's gut, but their removal does not interfere significantly with the termites' metabolism (Fig. 44).

The hypermastigid flagellates of termites do not encyst. Every time the growing insect moults, its infection is lost; in order to survive, it must somehow become re-infected from its neighbours (probably by ingesting flagellates in faeces taken directly from the anus of another termite). In the wood-eating roaches, the infection, once established, is permanent; some, at least, of the parasites of these insects do form cysts, by means of which other, newly hatched roaches are doubtless infected. Sexual reproduction occurs in many species of these flagellates; in those of at least one genus of wood-roach *(Cryptocercus)*, it has been shown that the onset of sexuality is related to the secretion by the host insect of moulting hormone. Two suborders are recognized in the order Hypermastigida: the

Lophomonadina, rather simpler organisms with only one anterior cluster of flagella, and the Trichonymphina, with two (rarely four) clusters of flagella. For further details of this fascinating group of organisms, including drawings, see Mackinnon and Hawes (1961, pp. 138–49) and the references cited therein: for a list of genera and some species, see Kudo (1966, pp. 480–92).

SUPERCLASS Opalinata

These organisms are the subject of much taxonomic confusion. They are large, oval, flattened Protozoa, covered with cilia, and possessing two or many similar nuclei. Almost all are parasites of the large intestine of frogs and toads (Amphibia, Salientia), a few having been described from other amphibia, reptiles and fish. They were for many years regarded as primitive ciliates (Protociliata), but now that electron microscopy has revealed the fundamental similarity between cilia and flagella many people regard the Opalinata as closer to the Mastigophora, or at least intermediate between them and the ciliates. Opalinata differ from ciliates as follows. (1) Their nuclei are not differentiated into macro- and micro-nuclei; (2) sexual reproduction involves the fusion of two dissimilar individuals (anisogametes), followed by encystment, and not conjugation as in ciliates; (3) asexual division is usually symmetrogenic (see p. 38), as in flagellates; however, homothetogenic fission also occurs. The problem is fully discussed in a monograph by Wessenberg (1961), which also contains a wealth of information on the genus *Opalina*. A justification of the taxonomic position of the group adopted in this book is given by Honigberg *et al.* (1964, footnote 8 on pp. 11–12).

5

PARASITIC AMOEBAE

The amoebae are members of the superclass Sarcodina of the subphylum Sarcomastigophora (see p. 18). All the parasitic forms, together with many non-parasitic ones (including the well-known *Amoeba proteus*), are grouped in the class Rhizopodea, subclass Lobosia, order Amoebida which consists of naked amoebae (i.e. without a shell or 'test') moving by means of (usually) lobose pseudopodia (p. 32). Only a few genera are obligate parasites, almost always of their host's intestine: these include *Entamoeba, Endolimax, Iodamoeba* and *Dientamoeba*,[1] all with one or more species inhabiting the gut of man, *Endamoeba* (whose name has been a constant source of confusion with *Entamoeba*) of insects, and a few others. Only two species (both of *Entamoeba*) are known to be harmful —*E. histolytica* of man and other mammals and *E. invadens* of reptiles. These sometimes invade the tissues of their hosts. Kudo (1966) includes all the obligate parasitic amoebae in a single family, the Endamoebidae,[2] and this, while simplifying taxonomy from the parasitologist's viewpoint, probably obscures their true relationship. It seems more likely that parasitism has been adopted independently by amoebae of at least two different groups. In addition to these, other normally free-living species of amoebae belonging to the families Hartmanellidae and Naegleriidae can invade the tissues of mammals, including man, and occasionally produce severe disease (see pp. 85–86 below).

[1] Some authors regard this genus as an aberrant flagellate of the family Rhizomastigida (pp. 64–65 above) which has lost the flagellate phase.
[2] With one exception—the little known genus *Paramoeba*, one species of which is a parasite of Chaetognatha: see Kudo (1966, p. 555).

All members of the order Amoebida lack any form of sexual reproduction, multiplying only by binary (occasionally multiple) fission. Many produce resistant cysts at certain stages of their life cycle, and all are phagotrophic (p. 34). The obligate parasitic forms are transmitted directly, as far as is known, usually by ingestion of faecal material containing cysts. Almost all parasitic amoebae produce cysts; two exceptions are *Entamoeba gingivalis,* which inhabits the mouth and upper pharynx of man, and *Dientamoeba fragilis,* of the large intestine of man. The former is presumably transmitted directly, the trophozoite (i.e. non-encysted stage) surviving since it does not have to pass through the stomach to reach its habitat. The transmission of *D. fragilis* is more puzzling, since the trophozoite is unlikely to survive a journey through the acid stomach contents even if it could live long enough outside one host to have a good chance of reaching a second. It is possible (though there is no direct evidence supporting this) that it travels from host to host as a passenger inside the eggs of a nematode, as other protozoa are known to do (e.g. *Histomonas,* p. 65). Apart from this hypothesis, no vectors are known for any parasitic amoebae, though cysts may be carried mechanically by insects such as flies and cockroaches which feed indiscriminately on faecal matter and on the food of man and other animals.

Genus *Entamoeba* Casagrandi and Barbagallo, 1895
This genus is distinguished by its nuclear structure (Fig. 45). The nucleoprotein is arranged in a peripheral ring (which appears granular in fixed specimens), lining the nuclear membrane, and in a small more or less central karyosome; the rest of the space within the nuclear membrane appears empty, and presumably contains fluid. All species except one are parasitic.

Levine (1961), following Hoare (1959), divides the species of this genus which infect man and domestic animals into four groups, based mainly on the number of nuclei present in the mature cyst (or, in one group, the absence of a cyst). The trophozoite of all species is uninucleate when it encysts, but in many species two or three nuclear divisions soon occur so that the mature cyst contains four or eight nuclei. In any sample a small number of immature cysts with less than the normal number of nuclei may be seen, together with occasional freaks having more. The cysts often possess, when young, a vacuole containing glycogen (presumably a food reserve) but this has usually disappeared (been metabolized) by the time the nuclear divisions (if any) are complete. Also, cysts sometimes contain structures called chromatoid bodies which in fresh preparations appear glass-like and stain intensely with various haematoxylins (hence their name). The chromatoid bodies, too, disappear as the cyst ages (though less rapidly than the glycogen). They are composed of ribonucleoprotein, arranged as ribosome-like particles in a regular, almost crystalline array (Pitelka, 1963). They presumably function as a reserve of this substance, and are used by the amoeba after it emerges from the cyst, when division and rapid growth occur. The disappearance of the chromatoid bodies as the cyst ages is more apparent than real—the ribonucleoprotein merely becoming dispersed throughout the cytoplasm (Neal, 1966). The shape of the chromatoid bodies is of some help in differentiating the groups of species. Levine's (1961) grouping of the species of this genus is as follows.

(1) *Cysts with 4 nuclei* and chromatoid bodies which are relatively broad rods, with blunt rounded ends (Figs. 47–9). Trophozoites and

---

Fig. 45. *Entamoeba histolytica*; trophozoite from smear of human faeces, with ingested erythrocytes and bacteria. ×1,500.

46. *E. histolytica*; trophozoite from culture, to show locomotion. ×1,500.

47. *E. histolytica*; uninucleate cyst from smear of human faeces. ×1,500.

48. *E. histolytica*; binucleate cyst from smear of human faeces. ×1,500.

49. *E. histolytica*; quadrinucleate cyst from smear of human faeces. ×1,500.

50. *E. hartmanni*; cyst from smear of human faeces. ×1,500.

51. *E. coli*; trophozoite from smear of human faeces. ×1,500.

52. *E. coli*; binucleate cyst from smear of human faeces. ×1,500.

53. *E. coli*; octonucleate cyst from smear of human faeces, containing chromatoid bodies. ×1,500.

54. *E. coli*; octonucleate cyst from smear of human faeces, without chromatoid bodies. ×1,500.

55. *Endolimax nana*; trophozoite from smear of human faeces. ×1,500.

56. *End. nana*; cyst from smear of human faeces. ×1,500.

57. *Iodamoeba buetschlii*; trophozoite from smear of human faeces. ×1,500.

58. *I. buetschlii*; cyst from smear of human faeces. ×1,500.

59. *Dientamoeba fragilis*; trophozoite from smear of human faeces. ×1,500.

60. 'Soil amoebae' (? *Naegleria* or *Hartmanella*) in section of human brain (right olfactory sulcus). ×1,500. (Section kindly provided by Drs M. Fowler and R. F. Carter.)

Abbreviations: B – bacterium; Ch – chromatoid body; Er – erythrocyte; G – glycogen vacuole; N – nucleus; P – parasite; Ps – pseudopodium; U – uroid.

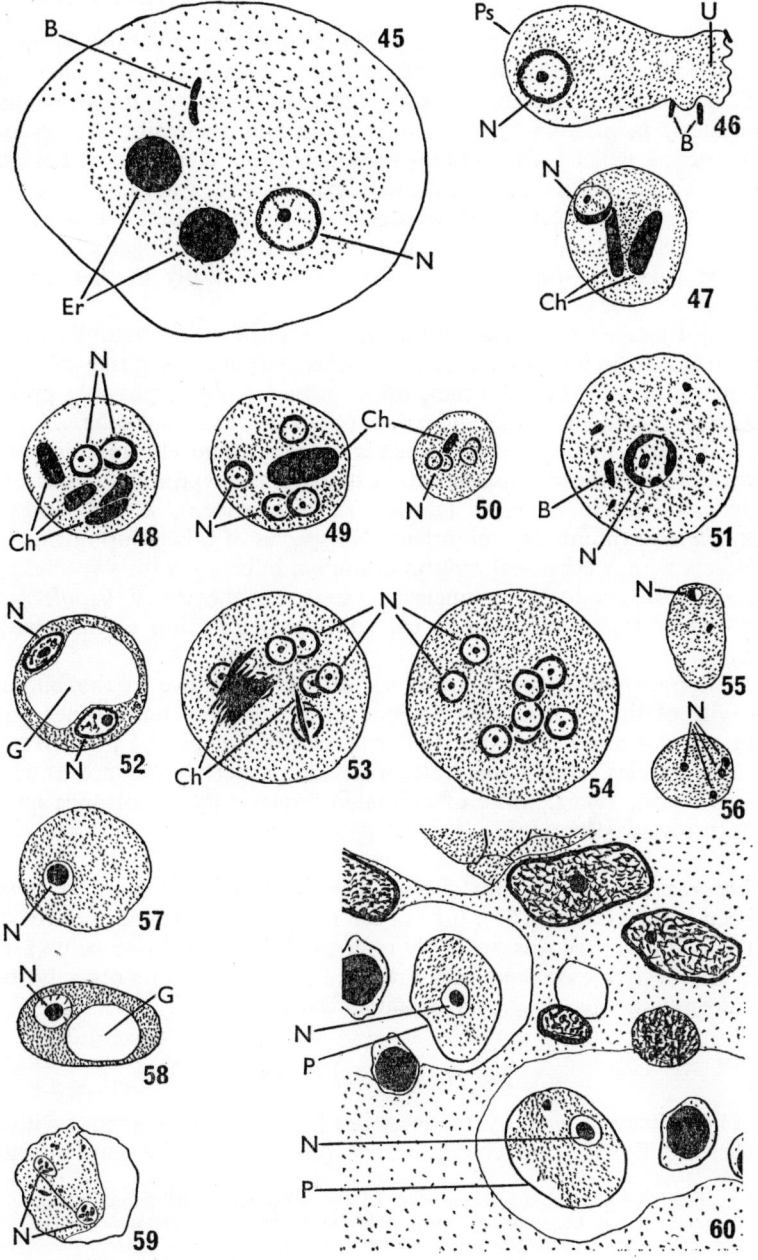

cysts of this group have nuclei of rather delicate structure. Species include *E. histolytica* (man, other primates, dogs, cats, pigs, rodents), *E. hartmanni* (man and possibly the other hosts of the previous species), *E. moshkovskii* (the only known non-parasitic species of this genus, which is found in sewage and is aptly referred to by Levine (1961) as 'a parasite... of the municipal digestive tract') and others. The first two species will be discussed in more detail below. *E. invadens* (reptiles), which is the only known pathogenic species apart from *E. histolytica,* belongs morphologically in this group also.

(2) *Cysts with 8 nuclei* and thin, splinter-like chromatoid bodies with pointed ends (Fig. 53). The nuclear structure is rather coarse. Species include *E. coli* (man, other primates, dogs, possibly pigs), *E. muris* (rats, mice), *E. gallinarum* (chickens, etc.) and others.

(3) *Cysts with 1 nucleus:* nuclear structure and chromatoids are variable, but the chromatoids are usually of the type seen in group 1. Species include *E. bovis* (cattle), *E. ovis* (sheep), *E. suis* (pigs), *E. chattoni* (monkeys) and others. No species of this group normally infects man; occasional reports of human infection with a species of *Entamoeba* producing uninucleate cysts (sometimes called '*E. polecki*') probably represent sporadic (or spurious[1]) infection of man with *E. suis.*

(4) *Species not producing cysts,* all of which live in the buccal cavity of their hosts. They include *E. gingivalis* of man, other primates, dogs and cats, and similar species of horses and pigs. Many other species parasitizing vertebrates and invertebrates are known (see Kudo, 1966): those of animals of veterinary importance have been reviewed by Hoare (1959).

Important species. E. histolytica. The hosts of this amoeba have been listed above (p. 80). It has been reported throughout the world, though many records probably refer to *E. hartmanni* (see below).

E. histolytica is sometimes pathogenic to man and its other hosts, causing amoebic dysentery or amoebiasis.

Morphology and life cycle

When alive and healthy, trophozoites (Figs. 45 and 46) move with a forward flowing movement, rather reminiscent of a garden slug

---

[1] An infection in which a resistant stage of a parasite passes unchanged through the alimentary canal of another animal, without undergoing development. Definite instances of this occurring in man are known among the coccidia (see p. 94).

(*Limax* sp.) and hence sometimes called 'limax-type movement'. The entire body, elongated and of fairly regular outline, flows smoothly forwards, all of its anterior end functioning as a single, broad pseudopodium. It is only at this anterior pole that the outer clear ectoplasm is sharply differentiated from the inner granular endoplasm. When kept on a slide at temperatures below 37°C, the amoeba becomes unhappy and moves in a far less single-minded fashion, putting out tentative pseudopodia in various directions and withdrawing them again. It does not survive long under such conditions, but soon rounds up and quietly dies. In this state, which is how it is usually seen in stained preparations on microscope slides, the distinction between ectoplasm and endoplasm is more clearly visible. The trophozoite contains a single nucleus of the type described above (p. 77), with rather fine karyosome and peripheral granules. Food vacuoles (but *not* a contractile vacuole) may be seen in the endoplasm, containing various objects such as bacteria, host-cell nuclei and—if the infection is pathogenic—host erythrocytes. Trophozoites are found in the host's large intestine, usually in the lumen where they measure 10–20 μ in diameter. Sometimes, for unknown reasons, trophozoites invade the mucosa and submucosa, and may spread to other tissues (chiefly the liver); in these richer environments they become larger (20–40 μ in diameter). This subject is discussed below under 'Pathogenesis'.

The trophozoites encyst only in the gut lumen—not in the tissues. They round up and become quiescent, and secrete a thin cyst wall. A diffuse mass of glycogen may be seen in young cysts, but this soon disappears. Chromatoid bodies (described on p. 78 above) are usually seen in young cysts, but they disperse as the cyst ages. The single nucleus of the trophozoite soon undergoes two divisions to produce the characteristic four of the mature cyst (Figs. 47–9) which on average measures about 12 μ in diameter (range 9·5–15·5 μ). The cyst is the transmissive stage; it does not develop further in the original host, but is passed out in the faeces. It can survive outside the host for some time, and when swallowed by another susceptible mammal passes unharmed through the stomach. In the small intestine (probably), the parasite emerges from the cyst (excysts) as a four-nucleate amoeba; its nuclei divide once more, and then cytoplasmic division occurs, producing eight small uninucleate amoebae which pass to the large intestine and grow to the full size. Here (and in the tissues, if invasion occurs) the amoebae multiply by binary fission.

Epidemiology

Infection with *E. histolytica* depends on ingestion of faecal material containing cysts, usually (since few people are coprophagous) in contaminated food or water. The source of infection is usually man, for though infected dogs or rats may occasionally pass the parasite to man it is probably more usual for them to become infected from him. Thus the infection is commoner in areas where food hygiene is less effective. Possible sources of infection are food-handlers who are themselves chronically infected without being ill ('cyst-passers'), and whose standards of cleanliness are not ideal. The use of untreated human faeces as a fertilizer is a common source of infection, as is contamination of water supplies and their inadequate purification. As mentioned earlier, coprophagous insects such as flies and cockroaches may carry viable cysts, either on their legs or in their intestine, to foodstuffs. The cysts will survive for several weeks outside the body if not desiccated. They are not killed by aqueous potassium permanganate, a 'disinfectant' often used for washing salads, etc., in tropical countries, but are readily killed by heat (50°C or above).

Pathogenesis

As mentioned above, *E. histolytica* usually lives in the gut lumen as a harmless commensal. Occasionally, for reasons which are unknown, the trophozoites will penetrate the mucosa and the muscularis mucosae and invade the sub-mucosa. Here they multiply and spread radially outwards below the mucosa to form a characteristically flask-shaped lesion, or ulcer, the centre of which is filled with cellular debris, lymphocytes, plasma cells and macrophages. Secondary bacterial infection may occur and then polymorphonuclear leucocytes will also be found. As the sub-mucosa is eroded by the amoebae (which are found chiefly at the advancing edges of the ulcer) many blood vessels are broken and the typical bloody dysentery results. Rarely, the ulcers perforate the gut wall entirely and cause peritonitis. A more common complication is the spread of amoebae via the blood vessels to other organs, where they also invade and destroy the tissue causing amoebic abscesses (in which bacteria are not present). The abscesses may become very large—several inches in diameter. The commonest site for their development is the liver, because most of the blood from the gut is carried there by the hepatic portal system; but they may be found in almost any organ, including lung and even brain.

Untreated amoebic dysentery may result in death from fluid and blood loss. If abscesses develop in the liver or elsewhere, apart from the damage done to the organ concerned, the abscess may finally rupture either through the body wall or into the peritoneal cavity with serious or fatal consequences.

Diagnosis

Amoebiasis is usually diagnosed by recovery of the parasites from faeces. If dysentery is present, active trophozoites may be seen in fresh faeces mixed with saline and examined microscopically at 37°C. (It is, however, possible for trophozoites of other, non-pathogenic amoebae to be passed out if diarrhoea is present due to some other cause, so any trophozoites seen have to be identified morphologically.) In diarrhoeic and normal faeces, cysts may be found either by direct microscopical examination or after their concentration from the sample. Finally, in cases of doubt, attempts may be made to cultivate the amoebae in suitable artificial media. All these methods are described in Chapter 12. For references to the use of serological methods in diagnosis see Neal (1966).

Treatment and prevention

Various drugs are available to control infection with *E. histolytica*, though complete cure is not always easily obtained. The alkaloid emetine has been used for many years (in various compounds) and is effective though rather toxic. A synthetic derivative, dehydroemetine, is apparently equally effective and less toxic. The antimalarial drug chloroquine is effective against amoebic abscesses in the liver but not elsewhere. Sometimes large abscesses have to be drained surgically.

Prevention of infection is entirely a matter of hygiene, both personal (washing of hands, avoiding the eating of raw vegetables and salads in dangerous areas, protection of food from coprophilic insects, etc.) and municipal (sewage disposal, water purification). If possible, food handlers in endemic areas should be examined for infection and treated if necessary (a counsel of perfection which is seldom practicable).

Other intestinal amoebae of man. Five other species of amoebae inhabit the human intestine (see Table 5). All are non-pathogenic, but when attempting to diagnose amoebiasis it is important to be

TABLE 5

INTESTINAL AMOEBAE OF MAN

	Trophozoite		Cyst			
	Size (μ)[1]	Nuclear structure	Size (μ)[1]	No. of nuclei (mature)	Chromatoid bodies	Special features
Entamoeba histolytica	10–40	Entamoeba type, delicate	9·5–15·5 (~12)	4	Broad, blunt ends	—
E. hartmanni	9–14	Entamoeba type, coarser	4–10·5 (~7·4)	4	Thin, sharp ends	—
E. coli	15–30	Entamoeba type, coarser	10–30 (~17)	8	None	—
Endolimax nana	6–12 (~8)	Not Entamoeba type	6–9 × 5–7	4	None	—
Iodamoeba buetschlii	5–20 (~11)	Not Entamoeba type	9–15	1	None	Persistent glycogen vacuole[2]
Dientamoeba fragilis	7–12	Not Entamoeba type (mostly binucleate)	No cyst			

[1] Means in parentheses.
[2] Stains golden-brown colour with iodine solution.

aware of their presence to avoid possible confusion. *Entamoeba hartmanni* closely resembles *E. histolytica*. The trophozoites are about the same size as the smaller individuals of *E. histolytica* (9–14 μ), while the cysts (Fig. 50), which have four nuclei and chromatoid bodies exactly like those of *E. histolytica*, are definitely smaller (4·0–10·5 μ in diameter, average 7·4 μ). Thus cysts of this type in human faeces which are less than 10 μ in diameter are almost certainly not *E. histolytica*. Much confusion surrounds the taxonomy of *E. histolytica* and *E. hartmanni*, the latter being often referred to as the 'small race' of the former; but the evidence that they are separate species seems to be good. It is probable that many records of '*E. histolytica*' indigenous to the temperate parts of the world refer to *E. hartmanni*. The remaining four species of human intestinal amoebae are listed in Table 5, together with their main distinguishing features (see also Figs. 51–9). Further details of these organisms and the parasitic amoebae of domestic animals are given by Levine (1961) and, of the latter group only, by Hoare (1959).

Genera *Hartmanella* Alexeieff, 1912 and *Naegleria* Alexeieff, 1912
These genera were thought until recently to be composed exclusively of amoebae living in damp soil or mud. However, symptomless infections of the throat of man with amoebae identified as *Hartmanella* sp.[1] have been found in about 1·7% of a group of 'normal' persons in the USA (Wang and Feldman, 1967). As most of the infections were in infants at the crawling stage, the likeliest means of infection was the ingestion of soil or dust containing cysts or trophozoites of the amoebae. It has been shown that, if similar amoebae are introduced experimentally into the nasal cavities of mice and monkeys, they may invade the mucosa and sometimes pass up the olfactory nerves to the olfactory bulbs and other parts of the brain, where they produce ulcers. If the inoculum enters the lungs, these, too, can be invaded. Amoebae can spread from these primary lesions to any organ in the body. The multiplying amoebae severely damage the tissues and experimental infections are often fatal.

Up to the end of 1968 there were 25 published reports of generalized fatal infections of man with amoebae which were not *E. histolytica*. One of these was found to be due to *Naegleria* sp. (which has a flagellate phase in its life cycle and is placed by Kudo (1966) in the family Naegleriidae) and the others may have been due

[1] The taxonomy of these amoebae is confused, and they are sometimes referred to as '*Acanthamoeba* sp.'. They are placed by Kudo (1966) in the family Amoebidae, but some workers classify them in a separate family, the Hartmanellidae.

to this genus or to *Hartmanella* sp. (Fig. 60). Some of the infected persons were known to have been swimming in muddy water and amoebae may have entered their mouths or noses in this way. References and further information on this interesting development in parasitology can be found in papers by Culbertson *et al.* (1966, 1968) and Červa and Novák (1968).

6

GREGARINES, COCCIDIA AND HAPLOSPOREA

In this and the next three chapters the large subphylum Sporozoa,[1] all members of which are parasitic, is considered. This subphylum is divided into three Classes—the Telosporea, Toxoplasmea and Haplosporea (a small and little-known group); some of the Telosporea (i.e. the suborder Haemosporina) and the Toxoplasmea are considered in Chapters 7 and 9 below. The Sporozoa are characterized by possessing, typically, simple spores which were originally a resistant stage involved in dispersion and transmission of the species. In more sophisticated members of the subphylum (e.g. the Haemosporina), the spore itself may be lacking, since with the adoption of an insect vector it is no longer necessary to protect the parasites while outside a host. In almost all Sporozoa, however, the infective stage (sporozoite) which develops within the spore of the more primitive members can be recognized (except in the Toxoplasmea, as far as is known). Other characteristics of the Sporozoa are the absence of cilia or flagella (except on the microgametes of some genera), and—in the majority of groups—the adoption of a type of multiple asexual fission called schizogony (see Chapter 2, p. 38).

CLASS Telosporea

The members of this class reproduce both sexually and asexually. They constitute the majority of species of Sporozoa, and include the causative organisms of malaria (which are considered in Chapter 7).

---

[1] Not everyone accepts the inclusion of the Piroplasms, discussed in Chapter 8, in the Sporozoa (see p. 117 below).

Probably all Telosporea are haploid for almost all of their life cycle, meiosis occurring in the first zygotic nuclear division (i.e. the first division following fertilization). Thus only the zygote is diploid. There are two subclasses, the Gregarinia and the Coccidia.

SUBCLASS Gregarinia

The gregarines are all parasites of the gut and haemocoel of invertebrates—mainly insects. For most of their life cycle they are extracellular, and they have no vector, transmission being by ingestion of the spores liberated from an infected host. The sporozoites released from the swallowed spores enter cells of the host (often those of the gut wall) where they increase in size. In most gregarines (the order Eugregarinida), the trophozoites ('gregarins') emerge from the host cell, but remain attached to it at first by the anterior end. In many eugregarines (the cephaline eugregarines) the body is divided into two compartments, an anterior protomerite and a posterior deutomerite (Fig. 61); on the front of the protomerite is a protrusion called the epimerite, by means of which the trophozoite is attached to the host cell. These two compartments should not be regarded as separate cells, since each trophozoite has only a single nucleus[1]—usually in the deutomerite. In other (acephaline) eugregarines this division of the body is not seen. The trophozoites eventually break away from the host cell, and move (in a way which is not fully understood; see p. 34) in the haemocoel, gut lumen or other body cavities of the host. In a minority of gregarines, the trophozoites remain intracellular and reproduce by schizogony (p. 38). Many such cycles of schizogony may occur, and each time a schizont matures a host cell is destroyed and many more are invaded by the released parasites (merozoites). For this reason gregarines of this type are often harmful to their hosts. Eventually, schizonts (often of a different type, with larger nuclei) are formed, the merozoites resulting from which remain extracellular and grow into sexual individuals or gametocytes ('sporadins'). In the gregarines without schizogony, all the trophozoites become gametocytes. Two gametocytes then become attached to one another and—after a longer or shorter interval—they encyst. This association of gametocytes (presumably of opposite sexes) is called syzygy (Fig. 62). The associated pair within the cyst (gametocyst) undergo nuclear division and bud off small (usually identical) individuals (gametes) from their surface. The gametes (which do not possess flagella) fuse in pairs to form zygotes, each of which then encysts within an oocyst (or 'spore'). Many zygotes are formed by the

[1] Kudo (1966) lists one possible exception.

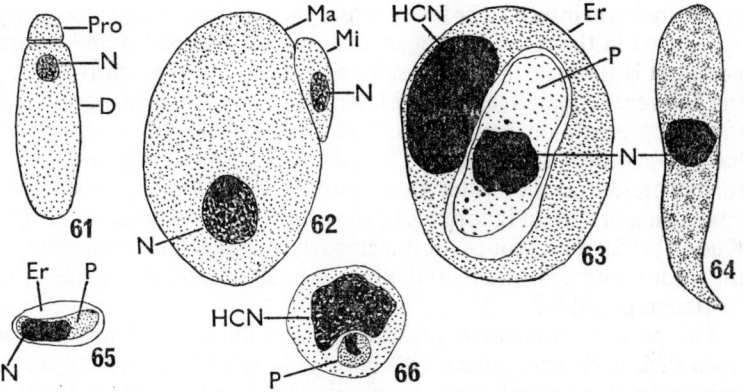

Fig. 61. *Gregarina ovata*; gregarine from an earwig. ×150.
62. *Adelea* sp.; gametocytes in syzygy (host unknown). ×750.
63. *Haemogregarina* sp. in erythrocyte of Mexican frog. ×1,500.
64. *Haemogregarina* sp. free in blood plasma of Mexican frog. ×1,500.
65. *Hepatozoon balfouri* in erythrocyte of jerboa. ×1,500.
66. *Lankesterella garnhami* in monocyte from smear of sparrow's spleen. ×1,500.

Abbreviations: D – deutomerite; Er – erythrocyte; HCN – host cell's nucleus; Ma – macrogametocyte (female); Mi – microgametocyte (male); N – nucleus; P – parasite; Pro – protomerite.

Eugregarinida, but fewer (often only two) by the gregarines which undergo schizogony. The oocysts are at first within the gametocyst but later, either before or after leaving the host, they escape from it. The contents of each sporocyst divide to form several (usually eight) sporozoites. The sporocysts are liberated in some way from the body of the host and, by virtue of their cyst wall, are resistant to desiccation and the other hazards of life outside a host. The sporocyst is the transmissive stage and, after it has been swallowed by a new host, the sporozoites emerge from it and the life cycle begins again.

Generally speaking, the gregarines which do not undergo schizogony are harmless to their hosts, while the others, since they increase greatly in number by repeated schizogonies within the host, are usually pathogenic (as mentioned above). Classification of the gregarines is based mainly on the presence or absence of schizogony. Following Grassé (1953), the Society of Protozoologists (Honigberg *et al.*, 1964) considers that the forms which undergo schizogony are not all closely related and so groups them in two separate orders— Archigregarinida (found mainly in marine annelids) and Neogregarinida (in insects); the forms which lack schizogony (the majority) are placed in the order Eugregarinida (found in annelids and arthropods). It is thought that the Archigregarinida are a primitive group, the Eugregarinida have evolved later, and have lost schizogony, and the Neogregarinida are the most recently developed group, which have re-acquired schizogony. This idea seems inherently improbable to the present author (though it must be admitted that his view is based on a profound ignorance of the group), and he prefers to follow Kudo (1966), who groups all the gregarines which lack schizogony in the order Eugregarinida[1] and those which possess it, in the order Schizogregarinida.

The parasite *Monocystis,* doubtless well known to many readers from their early encounters with biology in sixth form or first year at university, is an acephaline eugregarine. A good general account of the gregarines is given by Mackinnon & Hawes (1961), and a fuller account—in French—is given by Grassé (1953); Kudo (1966) lists many of the genera and species.

SUBCLASS Coccidia

Of the two orders within this class, the first, Protococcida, can be ignored. It is represented by only two species, parasitic in marine annelids. Kudo (1966) gives details. The second order, Eucoccida, contains three suborders: Adeleina, Eimeriina and Haemosporina (Chapter 7).

Order Eucoccida
Suborder Adeleina

The Adeleina are distinguished by the fact that males and females develop in association with each other (syzygy, as in gregarines; see p. 88). Many members of this suborder live in red or white blood cells, and sometimes other cells, of vertebrates (of all classes) and are called 'haemogregarines', an imprecise term which also includes some Eimeriina. These are transmitted from one vertebrate to another by invertebrate vectors (leeches, insects, ticks or mites), in which they undergo sexual reproduction; asexual multiplication (schizogony) occurs in the vertebrate host. The adeleine haemogregarines can be distinguished from malaria parasites (which also inhabit erythrocytes) by the fact that they contain no pigment. They are grouped in three families, Haemogregarinidae, Hepatozoidae and Karyolysidae (containing the genera *Haemogregarina*

---

[1] Kudo's (1966) ranking and orthography have been amended to comply with that of the classification proposed by the Society of Protozoologists.

(Figs. 63, 64), *Hepatozoon* (Fig. 65) and *Karyolysus* respectively). After schizogony in the vertebrate, the sexual precursors (gametocytes) are produced. These always inhabit blood cells, in which they are sucked up by the next blood-sucking invertebrate to feed on their host. If the invertebrate is of the correct species, the gametocytes associate in syzygy and produce sexual individuals (gametes). Fertilization occurs and the zygote, which in the Hepatozoidae is motile and is called an ookinete, usually encysts to form an oocyst in the lumen or wall of the vector's gut or in its haemocoel. In the Karyolysidae, the non-motile zygote divides into several motile sporokinetes which invade the vector's ovaries and encyst in the eggs to form sporocysts. The oocyst (or sporocyst) grows and its contents undergo repeated nuclear division and eventually differentiate into several (or many) small elongate uninucleate sporozoites (this process being called sporogony). The sporozoites by various routes enter the next vertebrate on which the vector feeds. In the species transmitted by leeches, the sporozoites may be injected when the leech feeds, but certainly in some species, and perhaps in all, they enter through the mucous membranes if the vertebrate eats the vector (or, in the Karyolysidae, its faeces, which contain sporocysts); the route of entry of most species is unknown. If this vertebrate is susceptible to the parasite, the sporozoites enter the appropriate cells and commence schizogony. Eventually gametocytes are produced and the cycle is complete.

This life cycle has many similarities to that of the malaria parasites (Chapter 7), except that syzygy does not occur in the latter. Its complexity may appear rather alarming at first sight, but basically it may be reduced to the following components: schizogony (asexual multiplication) in the vertebrate host, gametogony (development of the sexual gametes) begun in the vertebrate and completed in the invertebrate followed by fertilization in the invertebrate, and sporogony (multiplication immediately following fertilization) in the invertebrate host.

As far as is known, most adeleine haemogregarines are normally non-pathogenic; some species of the genus *Hepatozoon* may, however, cause disease in their hosts (e.g. *H. muris,* in mice). The remaining Adeleina, which do not inhabit blood cells and pass through a similar life cycle in a single host (vertebrate or invertebrate), are contained in the two families Adeleidae and Klossiellidae; the oocyst is passed out in the host's faeces or urine (compare the Eimeriidae, p. 97 below). More information about the Adeleina can be obtained from Kudo (1966).

Suborder Eimeriina

This group includes the organisms often colloquially referred to as 'coccidia', together with a few forms which, since certain stages inhabit the blood cells of vertebrates, are included in the loose term 'haemogregarines' (see p. 90 above). They are distinguished from the adeleines by the absence of syzygy. Apart from the blood-dwellers, most Eimeriina have only a single host, the oocyst being a thick-walled resistant stage which can survive for long periods outside the host and is the transmissive phase. They usually inhabit the cells of the intestinal epithelium of vertebrates (all classes), though some are found in deeper tissues and a few have been recorded from molluscs, arthropods and other invertebrates. The oocysts of most species enter and leave the host via the alimentary canal.

The eimeriine haemogregarines all belong to the genera *Lankesterella* and *Schellackia* (the latter is regarded by some as a synonym of *Lankesterella*). They are common in birds, reptiles and amphibia. *Lankesterella* (Fig. 66) is found in canaries, sparrows and other birds in this country, and may cause a fatal disease in nestlings (Lainson, 1959). Both schizogony and sexual development (including fertilization and sporogony) occur in various tissue cells of the vertebrate host. The sporozoites enter the blood cells (red or white or both), and are ingested by the invertebrate vector (leech or mite); they undergo no development in the vector and are carried passively to another vertebrate, to which they apparently gain access when it eats the vector (or at least crushes it in its mouth); they then enter the appropriate cells and commence schizogony.

Almost all the Eimeriina belong to the family Eimeriidae. (Pellérdy (1965) recognizes two other familes, Selenococcidiidae and Dobellidae, which will not be considered further here.) The classification of this family is much disputed. Pellérdy's (1965) version, which is based almost solely on the numbers of sporocysts and sporozoites within the oocyst (see below for explanation of these terms), is followed here, although it is probably too artificial: he recognizes 25 genera, grouped into 8 subfamilies (see Table 6). Substantially the same 'mathematical' division into genera is presented diagrammatically by Levine (1961, p. 159).

Most species in the family Eimeriidae belong to the two genera *Eimeria* and *Isospora*. *Eimeria* includes many important parasites of domestic mammals and birds, while *Isospora* is the only member of the suborder Eimeriina to parasitize man (two species, *I. belli* and *I. hominis,* are rare but mildly pathogenic inhabitants of the human

TABLE 6
SUBFAMILIES AND GENERA OF THE EIMERIIDAE
(after Pellérdy, 1965)

Subfamily	Genus	Number of	
		Sporocysts	Sporozoites[1]
Cryptosporidiinae	Cryptosporidium	0	4
	Pfeiferinella	0	8
	Schellackia	0	8
	Tyzzeria	0	8
	Lankesterella	0	∞
Caryosporinae	Mantonella	1	4
	Caryospora	1	8
Cyclosporinae	Cyclospora	2	2
	Isospora	2	4
	Dorisiella	2	8
Eimeriinae	Eimeria	4	2
	Wenyonella	4	4
	Angeiocystis	4	8
Yakimovellinae	Octosporella	8	2
	Yakimovella	8	∞
Pythonellinae	Hoarella	16	2
	Pythonella	16	4
Barrouxinae	Barrouxia	∞	1
	Echinospora	∞	1
Aggregatinae	Merocystis	∞	2
	Pseudoklossia	∞	2
	Aggregata	∞	3
	Caryotropha	∞	12
	Myriospora	∞	∞
	Ovivora	∞	∞

[1] The number of sporozoites is that in each sporocyst except in the Cryptosporidiinae, when the total number per oocyst is given.

intestine). The species of veterinary and medical importance are well discussed by Levine (1961) and Davies, Joyner and Kendall (1963).

Genus *Eimeria* Schneider, 1895

This genus contains a large number of species parasitizing vertebrate animals of all classes. A few species have been recorded, some rather doubtfully, from annelids, arthropods (mostly centipedes) and protochordates (see Pellérdy, 1965). None infects man, though oocysts of one or two species parasitizing fish have been seen in the faeces of persons who have eaten infected fish and have been erroneously described as parasites of man. Some of the more important species of domestic animals are listed in Table 7.

The life cycle is typical of the group. An oocyst (Fig. 67), when swallowed, hatches in the small intestine of the host (probably under the influence of mechanical pressure, pepsin and trypsin); the sporozoites emerge from the sporocysts, and penetrate the cells of the intestinal mucosa. They may round up and grow in these cells, or they may be carried by macrophages elsewhere in the body, depending on the species. The growing forms are called trophozoites (Fig. 68). Most of them begin nuclear division, thus becoming schizonts (Fig. 69). Cytoplasmic division then occurs, to produce the small (usually about 5–15 μ long by 1–2 μ wide), uninucleate organisms called merozoites. (Note that this process, schizogony, is asexual.) The size of the schizonts varies widely with different species, from about 10 μ to several hundred microns in diameter. The number of merozoites produced varies from about 16 up to thousands. These quantities may also vary at different stages in the life cycle of the same species. The merozoites are liberated from the host cell when it and the schizont rupture, and, in most species, re-enter other cells (either nearby or, in some species, in more distant tissues) to recommence schizogony for a varying, but limited, number of generations. Sooner or later, for reasons which are not known, the merozoites do not recommence schizogony but instead enter host cells (usually of the intestinal mucosa) and develop by a process called gametogony into sexual individuals, the gametocytes (Fig. 70). These grow: the female or macrogametocyte remains uninuclear, but the male microgametocyte undergoes repeated nuclear division and finally produces at its surface a large number of small (2–3 μ long), curved organisms consisting mainly of nucleus, mitochondrion and three flagella. These small organisms are the motile male gametes, or microgametes. They swim away in search of a female and, on finding one, one of them fertilizes her. The fertilized female, now called a zygote, encysts (still within its host cell). Within the thick cyst wall, its nucleus (now diploid, as a result of fertilization) divides meiotically; cytoplasmic division follows, producing four cells called sporoblasts. Each sporoblast then encysts (within the oocyst) to form a sporocyst, the contents (nucleus and cytoplasm) of which divide again to produce the two sporozoites (Fig. 67). Often some residual cytoplasm is left unused after either or both of these divsions. Since the first post-fertilization nuclear division is a meiosis, it is clear

TABLE 7

SOME OF THE MORE IMPORTANT SPECIES OF *Eimeria* AND *Isospora*
WHICH INFECT DOMESTIC ANIMALS
(for descriptions see Davies, Joyner and Kendall, 1963)

Host	Species	Habit at	Pathogenicity
Cattle	E. bovis	Small & large intestine	Fairly high
	E. zurnii (+17 other species)	Small & large intestine	Fairly high
Sheep and Goats	E. ahsata	?	Moderate
	E. arloingi (+8 other species)	Small intestine	Moderate
Pigs	E. debliecki (+5 other species)	Small & large intestine	Mild-moderate
	I. suis	Small intestine	Mild
Chickens	E. necatrix	Small intestine & caeca	High
	E. tenella (+6 other species)	Caeca	High
Turkeys	E. adenoeides	Small & large intestine	Moderate
	E. meleagrimitis (+5 other species)	Small intestine	Moderate
Ducks	E. bucephalae (+1 other species[1])	Small intestine	High
Geese	E. truncata (+5 other species)	Kidney tubules	High
Rabbits	E. stiedae	Bile duct of liver	High
	E. irresidua	Small intestine	Moderate
	E. magna (+2 or more other species)	Small intestine	Moderate
Dogs and Cats	E. canis (dog only)	?	Moderate
	I. bigemina	Small intestine	High
	I. felis (+1 or 2 other species of Eimeria in cats only, and 1 other species of Isospora)	Small intestine (rarely large)	Moderate

[1] A species of another genus, *Tyzzeria perniciosa*, is pathogenic to ducks.

that species of *Eimeria* (and indeed as far as is known all Eimeriina as well as Haemosporina, and, probably, Adeleina) are haploid for almost all of their life cycle. At some stage during the differentiation

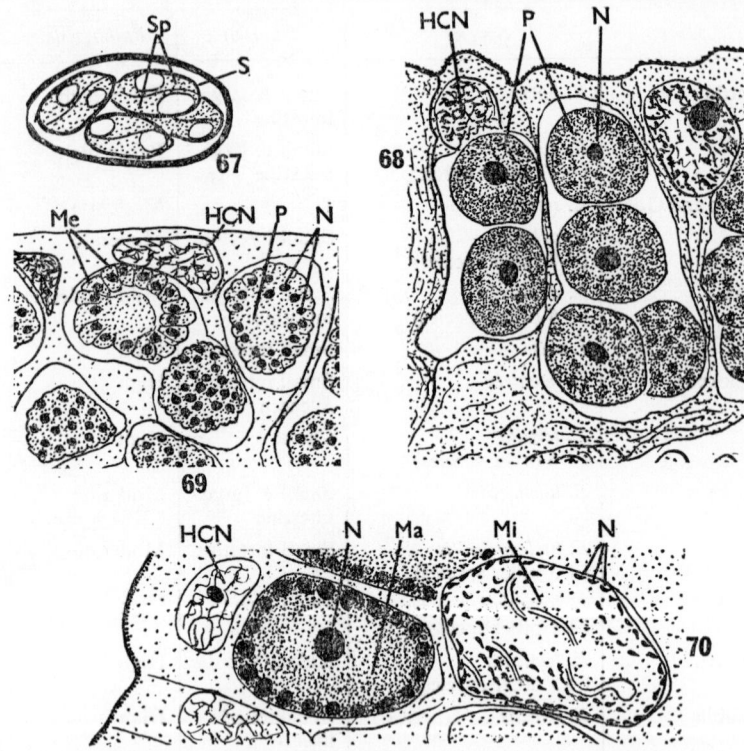

Fig. 67. *Eimeria stiedae*; oocyst in rabbit's faeces (preserved). ×1,500.
68. *E. perforans*; trophozoites in section of rabbit's intestine. ×1,500.
69. *E. perforans*; schizonts in section of rabbit's intestine. ×1,500.
70. *E. perforans*; gametocytes in section of rabbit's intestine. ×1,500.

Abbreviations: HCN – host cell's nucleus; Ma – macrogametocyte (female); Me – merozoite; Mi – microgametocyte (male); N – nucleus; P – parasite; S – sporocyst; Sp – sporozoite.

of the oocyst contents (sporogony), the oocyst leaves its host cell and passes down the host's intestine to the outside world. There it can survive for a considerable time, until eaten by another susceptible host and thus enabled to re-start the complicated life cycle (see Fig.

Gregarines, Coccidia and Haplosporea

71). This developmental cycle has many similarities with that of the Haemosporina (see Chapter 7), as well as with that of the adeleine haemogregarines (see above), except that in the latter two groups

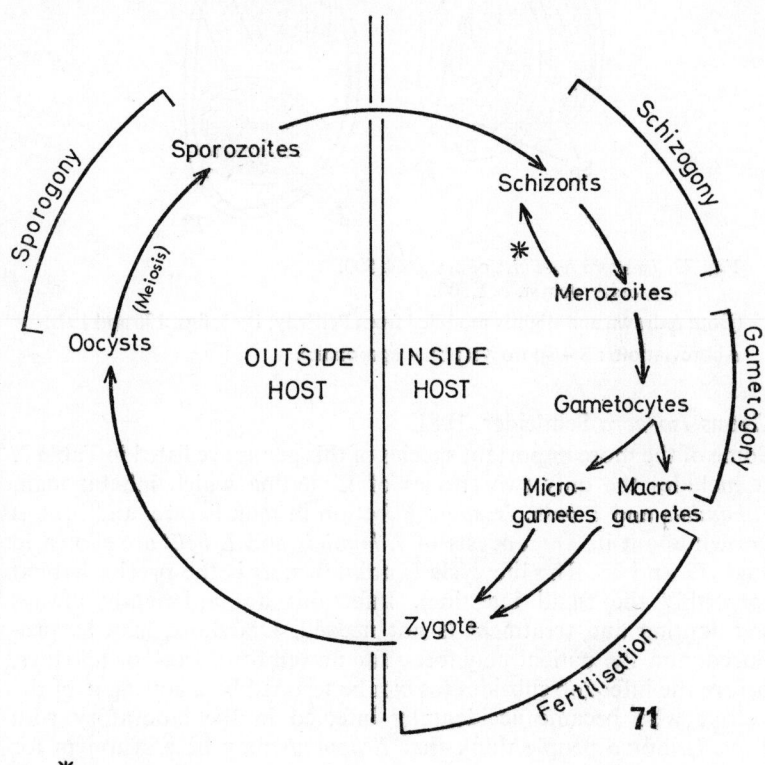

* Not in all species

Fig. 71. Diagrammatic representation of the life cycle in the suborder Eimeriina (except 'haemogregarines').

fertilization and sporogony occur in a second, vector host and at no time is the parasite exposed to the dangers of the outside world.

All the Eimeriidae have life cycles very similar to this, differing only in minor ways such as the numbers of sporocysts and sporozoites produced (see Table 6), the numbers of cycles of schizogony, and the location of the various developmental stages in the body of the host.

D

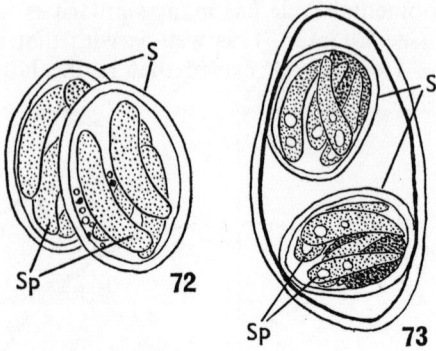

Fig. 72. *Isospora hominis*; oocyst. ×1,500.
73. *I. belli;* oocyst. ×1,500.
(Both redrawn and slightly modified from Pellérdy, 1965, figs. 196 and 197.)
Abbreviations: S – sporocyst; Sp – sporozoite.

Genus *Isospora* Schneider, 1881
Some of the more important species of this genus are listed in Table 7. It includes the only two species of Eimeriina which inhabit man, *I. hominis* and *I. belli*. *Isospora* infection in man is rare, and little is known about it. The oocysts of *I. hominis* and *I. belli* are shown in Figs. 72 and 73. The life cycle is unknown, as is the precise habitat (probably the small intestine). Infections are apparently always self-limiting and treatment is not needed. Diarrhoea may be produced, and the patient may feel quite unwell for a week or ten days, before the infection subsides (as can be testified by a colleague of the author who became accidentally infected in the laboratory with *I. belli*). Some people think that *I. hominis* may be a synonym for *I. bigemina* of dogs and cats from which, if this is true, most human infections presumably arise.

CLASS Haplosporea
The correct taxonomic position of this small and little-known group is uncertain, but as they produce simple spores and undergo schizogony (of a kind) they are provisionally placed in the Sporozoa; they show some affinities also with the Cnidospora (Chapter 10), and may possibly be the descendants of an erstwhile 'missing link' between the ancestors of the Sporozoa and those of the Cnidospora.

The Haplosporea are parasites of invertebrates and fishes. In the connective tissue or haemocoel of the host they undergo (extracellularly) a type of schizogony resulting in the formation of binucleate cells, which, by a process which is called 'sporogony' but is apparently not homologous with sporogony of the Telosporea since sex is not involved, differentiate into spores. The spore is the infective phase. When one is swallowed by a new host, a small, amoeboid organism emerges from it and migrates to its site of development. Kudo (1966) describes several genera and species of Haplosporea.

7

MALARIA PARASITES AND THEIR RELATIVES

The term 'malaria parasites' is generally restricted to species of the genus *Plasmodium,* which parasitize reptiles, birds and mammals. *Plasmodium* is classified in a family of its own, the Plasmodiidae. Closely related genera, which do not infect man, are usually grouped into two separate families, the Haemoproteidae and the Leucocytozoidae; some authors, however, consider that they should be united in a single family, the Haemoproteidae. All of these families are members of the suborder Haemosporina within the sub-phylum Sporozoa (see Chapter 1). They are obligate intracellular parasites for almost all of their life cycle, and have two hosts: a vertebrate in which asexual reproduction or schizogony[1] occurs, and an invertebrate (always a blood-sucking dipterous insect) in which sexual reproduction occurs, the insect host being regarded as the vector (see p. 13). References and further details concerning all the parasites mentioned in this chapter can be found in the monograph by Garnham (1966c).

Family Plasmodiidae
This family consists of the single genus *Plasmodium* Marchiafava and Celli, 1885, a genus of Sporozoa which undergoes sexual reproduction in mosquitoes (Diptera, Culicidae) and asexual reproduction (schizogony) in vertebrates. In the latter host, schizogony occurs in certain fixed tissue cells and also in erythrocytes; gametocytes develop

[1] Schizogony (see Chapter 2) is a type of asexual reproduction resulting in the production of four or more progeny (merozoites), during which all or almost all the nuclear divisions are completed before cytoplasmic cleavage begins.

in the erythrocytes; the intra-erythrocytic forms metabolize haemoglobin, producing a characteristic yellow, brown or black 'malarial pigment' or haemozoin, a compound containing haeme, in vacuoles within their cytoplasm. It has been suggested that the genus should be split into three or more, but such schizoid tendencies are now usually restricted to the subgeneric level. The following subgenera are generally recognized:

(1) In primates: *Plasmodium* and *Laverania* (transmitted by mosquitoes of the genus *Anopheles* only).

(2) In lemurs and lower mammals: *Vinckeia* (transmitted only by *Anopheles*).

(3) In birds: *Haemamoeba, Huffia, Giovannolaia* and *Novyella* (transmitted by various genera of mosquitoes, anopheline and culicine).

(4) In reptiles: *Sauramoeba, Carinamoeba* and *Ophidiella* (transmission unknown).

The life cycles of members of all these subgenera are very similar, and indeed have many similarities with those of the suborder Eimeriina (Chapter 6) (cf. Figs. 71 and 86). The vertebrate host is infected by means of small ($10-15 \times 0.5-1$ μ) fusiform, uninucleate sporozoites (Fig. 74), injected in the saliva of an infected mosquito when the latter feeds.[1] The sporozoites penetrate fixed tissue cells and commence schizogony. The cells in which this occurs are different in the different groups of subgenera. In those parasitizing mammals they are always (and only) liver parenchyma cells (Fig. 75): in those parasitizing birds and reptiles, the first generation (primary exoerythrocytic schizonts) is in lymphoid-macrophage cells in the skin near the site of the mosquito's bite and subsequent (secondary exoerythrocytic) schizonts are found in various lymphoid-macrophage cells throughout the body.[2]

In the former group, most (or all) of the products of division (merozoites) of the primary exoerythrocytic schizonts enter erythrocytes. At first they are seen as small ring-shaped parasites (Fig. 76); they then grow (during which process they are known as trophozoites; see Fig. 77); and finally they commence nuclear division as the beginning of erythrocytic schizogony (Fig. 78). The 'classical' view is that, in all species except *P. falciparum,* other merozoites re-enter

[1] As only female mosquitoes feed on blood, only they can serve as vector hosts of malaria parasites.
[2] The terminology of these stages is rather confused: the first generation schizonts, not in erythrocytes, are best called primary exoerythrocytic (=pre-erythrocytic=cryptozoic) schizonts; subsequent generations are called secondary exoerythrocytic (=exoerythrocytic=metacryptozoic=phanerozoic) schizonts.

liver cells and develop as successive generations of secondary exoerythrocytic schizonts, but this has never been conclusively demonstrated and is now doubted by some authorities (Garnham, 1967); it may rather be that some primary exoerythrocytic schizonts have a dormant phase, or grow much more slowly. In the subgenera which infect birds and reptiles, there is no doubt that some of the merozoites resulting from the primary exoerythrocytic schizonts invade other lymphoid-macrophage cells to produce secondary exoerythrocytic schizonts. Some of the merozoites produced by these secondary schizonts invade other lymphoid-macrophage cells, while others enter red blood cells (or their precursors) to initiate erythrocytic schizogony, both types of schizogony continuing side by side. Merozoites from erythrocytic schizonts of mammalian malaria parasites invade only other erythrocytes, but those of avian and reptilian malarias may enter either other red-blood cells or lymphoid-macrophage cells; in the latter they develop as secondary exoerythrocytic schizonts. Some of the merozoites entering erythrocytes (in both groups) develop, not as erythrocytic schizonts, but as gametocytes (sexual individuals; see Figs. 81, 82): one of the great unsolved problems in the Eucoccida as a whole is the elucidation of the factors determining whether a merozoite behaves as a male, female or asexual individual. The male and female gametocytes (which can be differentiated in stained blood films as the latter have a more compact nucleus and very basophilic cytoplasm) do not divide, but remain within their host erythrocytes until they either die or are ingested by a mosquito in which they continue their development in the insect's stomach. The female emerges from the red cell, rounds up (if not already spherical) and, now a mature gamete, awaits the arrival of a male. The male gametocyte meanwhile becomes very active: emerging from its host cell, it undergoes three nuclear divisions and develops eight flagella, all in the space of about 15–20 minutes; the flagella emerge from the surface of the gametocyte (this process is termed 'exflagellation'; see Fig. 83), one nucleus passes inside the membrane surrounding the axoneme of each flagellum, and the resulting male gametes break off the gametocyte and swim rapidly away: those which meet female gametes, penetrate and fertilize them. The process of gamete-formation, of both sexes, is called gametogony. As in the Eimeriidae (Chapter 6), the females are termed macrogametocytes and macrogametes, while the males are often spoken of as microgametocytes and microgametes. The fertilized female elongates into a motile ookinete (Fig. 84) (the male and female nuclei fuse at this stage or shortly after), which burrows into

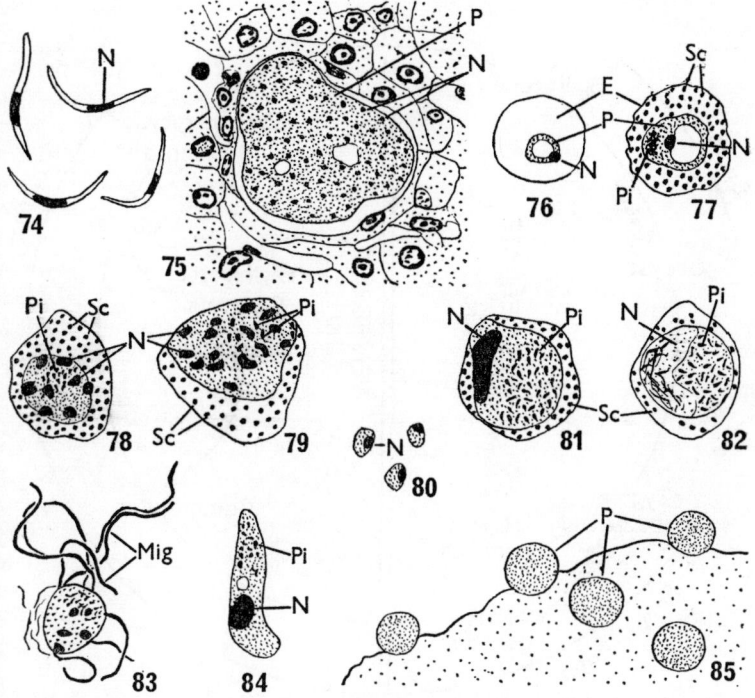

Figs. 74–85. *Plasmodium vivax* (except Fig. 74, which is *P. cynomolgi*). All ×1,500 (except Fig. 85).

74. Sporozoites from salivary glands of anopheline mosquito.
75. Primary exoerythrocytic schizont in section of human liver.
76. Young trophozoite ('ring form') in human red blood cell.
77. Trophozoite in human red blood cell.
78. Young schizont in human red blood cell.
79. Older schizont in human red blood cell.
80. Liberated merozoites in human blood plasma.
81. Macrogametocyte (female) in human red blood cell.
82. Microgametocyte (male) in human red blood cell.
83. Development of male gametes from microgametocyte ('exflagellation'). (Slide made from preparation in vitro.)
84. Ookinete from mid-gut of anopheline mosquito.
85. Oocysts on mid-gut of anopheline mosquito (mounted whole). ×150.

Abbreviations: E – erythrocyte; Mig – microgamete (male); N – nucleus; P – parasite; Pi – pigment; Sc – Schüffner's dots.

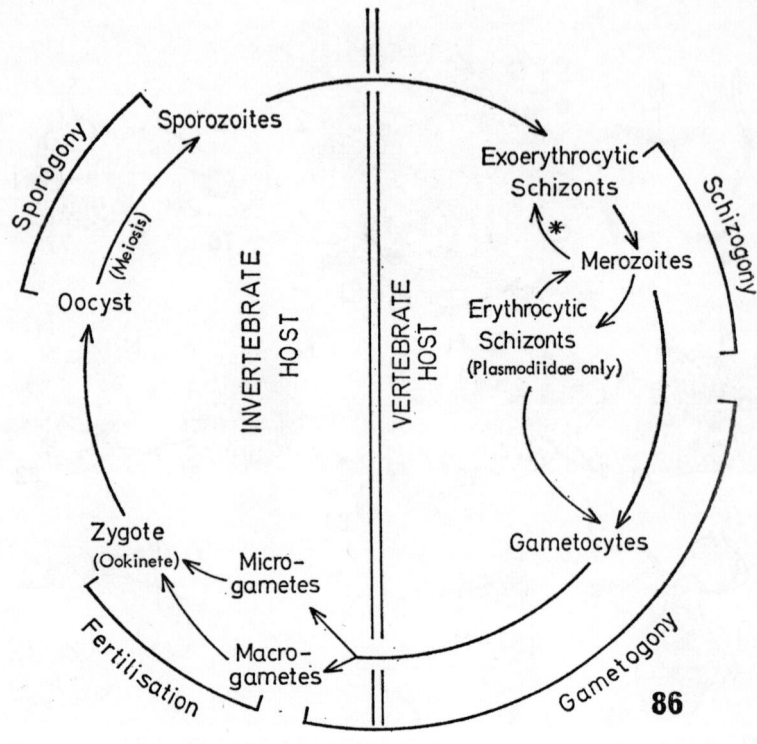

* Not in *Plasmodium falciparum*

Fig. 86. Diagrammatic representation of the life cycle in the suborder Haemosporina.

and, usually, through a cell of the single-layered epithelium forming the wall of the mosquito's stomach. On the outer surface of the stomach wall it encysts and becomes an oocyst (Fig. 85); its contents then undergo sporogony, a similar process to schizogony, which results in the production of thousands of uninucleate sporozoites (Fig. 74). The first nuclear division of sporogony is a reduction division and all stages except the ookinete are haploid. The oocyst, which has been increasing in size throughout sporogony, bursts and the liberated sporozoites spread throughout the mosquito's haemocoel. Most of them eventually penetrate the salivary glands, where

Malaria parasites and their relatives

they remain until the insect has its next blood meal and they are injected with the saliva into the animal on which it is feeding. If the latter is a susceptible host, they will then enter the appropriate tissue cells and commence primary exoerythrocytic schizogony. This complicated life cycle is summarized diagrammatically in Fig. 86. It is worth noting that the development in the mosquito appears to have no harmful effect on the insect.

MALARIA IN MAN

Four species of *Plasmodium* have man as their main (or only) vertebrate host. Their geographical distribution, and main distinguishing characteristics, are summarized in Table 8. All are transmitted by mosquitoes of the genus *Anopheles*.

Plasmodium (Plasmodium) vivax (Figs. 75–85). This species is the most widespread of the human malaria parasites, and used to be common in southern England. (Oliver Cromwell is said to have suffered from it.) It existed precariously in this country, where the winters are too cold for adult mosquitoes and so its transmission occurred only in the summer; for the rest of the year it survived in its vertebrate hosts. Increasing urbanization and drainage of swamps, etc., which reduced the contact between *Anopheles* and man, probably led to its extinction on this island, though a few indigenous infections were recorded after the 1914–18 war, the mosquitoes deriving their infections from returning soldiers. Subsequently malaria eradication programmes have removed it from many of its other temperate habitats. It produces a relatively mild disease (benign tertian[1] malaria) in man, which, though debilitating, is not fatal. If untreated, it persists for many years, relapses eight years after infection having been recorded, presumably due to the successful reinvasion of erythrocytes by merozoites produced by persisting exoerythrocytic schizonts in the liver. One of the most characteristic features of this parasite, as seen in thin blood films, is the effect it produces on its host erythrocyte. As the parasite grows, the red cell becomes enlarged (to a much greater extent that could be due simply to mechanical pressure from the parasite), pale in

---

[1] The adjective tertian is derived from the fact that the characteristic periodic fevers, which are associated with the rupture of erythrocytic schizonts (see p. 110 below), occur usually at intervals of 48 hours: thus, if the day of the first fever is numbered 1, the next fever will occur on day 3. The adjective benign, which would probably be disputed by those who have suffered from the disease, refers to the fact that this type of malaria does not result in death of the patient.

TABLE 8
CHARACTERISTICS OF SPECIES OF *Plasmodium* WHICH INFECT MAN

Species	Geographical distribution	Duration of schizogony		No. of merozoites per schizont	
		Exoerythro-cytic	Erythro-cytic	Exoerythro-cytic	Erythro-cytic
P.(P.) vivax	Worldwide, in tropical, sub-tropical and warmer temperate regions	8 days	48 hours	~10,000	12–24
P.(P.) malariae	Worldwide but scattered, mainly tropical and subtropical	14–15 days	72 hours	~15,000	6–12
P.(P.) ovale	Tropical Africa; also occasionally in other parts of tropics and subtropics (possibly of exogenous origin)	9 days	48 hours	~15,000	6–12 (12–24 in relapses)
P.(L.) falciparum	Worldwide, in tropics, subtropics and warmer temperate regions	5½ days	48 hours	~30,000	8–24

colour (probably due to the ingestion and digestion of its haemoglobin by the parasite) and its surface membrane becomes covered with closely packed, fine dots, which stain pink with Giemsa's stain. The dots, named after their discoverer as Schüffner's dots, are not seen in unstained parasitized cells and neither has electron microscopy so far revealed their nature.[1] Work with fluorescent antibodies on other species of malaria has suggested that they may represent aggregations of some antigenic material, perhaps excreted by the

[1] See, however, p. 110

TABLE 8 (continued)

Main morphological characteristics of erythrocytic forms

'Ring' form	Trophozoite	Schizont	Gametocytes	Host erythrocyte[1]
At least one-third diameter of erythrocyte	Amoeboid	10 μ diameter	Round or ovoid, male: 9 μ, female: 10–11 μ	Enlarged, stippled ('Schüffner's dots')
At least one-third diameter of erythrocyte	Compact, often bandlike	7 μ diameter	Round or ovoid, 7 μ	Not enlarged; very faint ('Ziemann's') stippling after prolonged staining only
At least one-third diameter of erythrocyte	Compact	7 μ (larger in relapses)	Round or ovoid, 9 μ	Slightly enlarged, stippled ('Schüffner's dots'); may be distorted and elongated
Very small at first; 2 nuclei commonly; some apparently on edge of cell (accolé forms)	Compact; rarely seen in peripheral blood	5 μ; rarely seen in peripheral blood	Crescentic, male: 9–11 μ long, female: 12–14 μ	Not enlarged; often with fewer larger dots ('Maurer's clefts')

[1] These alterations do not develop until some growth of the parasite has occurred.

parasite and deposited on the red cell's plasmalemma; but no definite evidence as to their nature has yet been obtained, and they remain enigmatic.

The parasite owes its specific name ('the lively *Plasmodium*') to the active amoeboid movement shown by the growing trophozoite; thus in dried, stained blood films the trophozoites may be very irregular in shape.

Plasmodium (Plasmodium) malariae (Figs. 87–92). This species is less common than *P. vivax*. It holds the current record for longevity of infection in man, 40 years or more in untreated persons. Like *P. vivax*, it is not a very pathogenic organism, although chronic infections sometimes give rise to a lethal kidney condition. *P. malariae* is the only species of those commonly infecting man which can also infect other primates. It is found in the West African chimpanzee.[1] Since man and chimpanzees seldom live in close proximity, the importance of the latter as a reservoir of human disease is probably negligible. Table 8 shows that both intra- and exoerythrocytic schizogony are slower in this species than the others, and this is true also of sporogony: at 24°C, the latter cycle takes about 21 days in mosquitoes, compared with about 16 for *P. ovale*, 11 for *P. falciparum* and 9 for *P. vivax*. *P. malariae* does not have as great an effect upon its host erythrocyte as does *P. vivax*. The red cell is not enlarged, and normally no stippling is seen. However, prolonged staining reveals very fine dots scattered somewhat irregularly over the plasmalemma (Ziemann's dots).

Plasmodium (Plasmodium) ovale (Figs. 93–7). This, the rarest of the four species which infect man, was not finally recognized as a distinct species until 1922, and not all workers accepted the distinction even then. Morphologically, it shows some resemblance to *P. malariae* (though the duration of erythrocytic schizogony is different) while its effect on its host cell resembles that of *P. vivax*. It has been well described as '*P. malariae* in a *P. vivax*-type cell'. The host cell shows pronounced Schüffner's dots and is enlarged, though slightly less so than is that infected with *P. vivax*; it seems to become unusually pliable, so that, in the making of a thin film, it may be drawn out into an elongated oval shape (hence the specific name) and even show a tattered and torn appearance at one end (Fig. 94). *P. ovale*, like the two preceding species, is not greatly pathogenic.

Plasmodium (Laverania) falciparum (Figs. 98–107). This parasite is probably the commonest species of *Plasmodium* to infect man. It also has the doubtful distinction of being 'almost unchallenged in its supremacy as the greatest killer of the human race over most parts of Africa and elsewhere in the tropics' (Garnham, 1966c). The pathology of the disease (malignant tertian malaria) caused by it is described below (pp. 110–112). The specific name is derived from the fact that, in this species, the gametocytes are crescentic (Figs. 106, 107).

---

[1] The other species which infect man can develop exoerythrocytically in the liver of chimpanzees, but the blood is more or less resistant to infection unless the animal has its spleen removed.

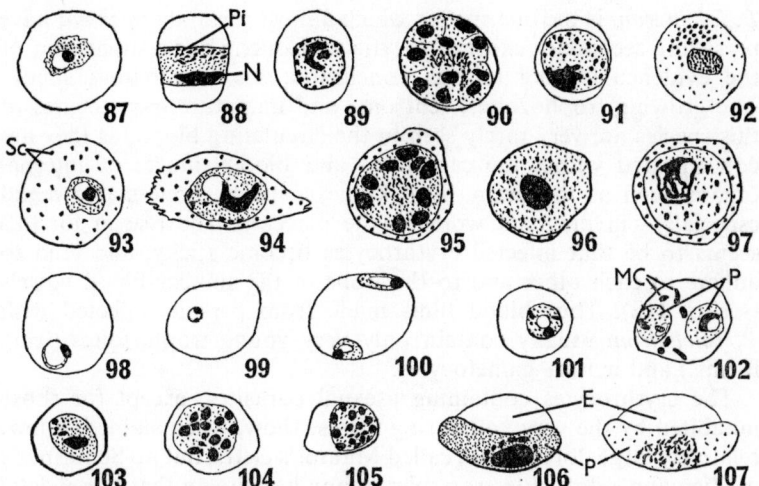

Figs. 87–92. *P. malariae* in human red blood cells. All ×1,500.

    87. Young trophozoite ('ring-form').
    88. Trophozoite ('band-form').
    89. Trophozoite (compact form).
    90. Schizont.
    91. Probable macrogametocyte (female).
    92. Microgametocyte (male).

Figs. 93–97. *P. ovale* in human red blood cells. All ×1,500.

    93. Young trophozoite ('ring-form').
    94. Trophozoite (note characteristic distortion of blood cell).
    95. Schizont.
    96. Macrogametocyte (female).
    97. Microgametocyte (male).

Figs. 98–107. *P. falciparum* in human red blood cells. All × 1,500.

    98. Young trophozoite ('ring form').
    99. Very small trophozoite ('ring') with 'double' nucleus.
    100. Two 'ring forms', marginal in position ('accolé forms').
    101. Older trophozoite (peripheral blood; slide kindly provided by Mr L. Hanning).
    102. Two older trophozoites (peripheral blood).
    103. Older trophozoite (peripheral blood).
    104. Schizont (placental blood).
    105. Schizont (peripheral blood; slide kindly provided by Mr L. Hanning).
    106. Macrogametocyte (female).
    107. Microgametocyte (male).

Abbreviations: E – erythrocyte; MC – Maurer's clefts; N – nucleus; P – parasite; Pi – pigment; Sc – Schüffner's dots.

P. falciparum is the one species which almost certainly does *not* have persisting secondary exoerythrocytic schizonts. (For a discussion of the evidence leading to this conclusion, see Garnham (1966c).) The growing trophozoites, schizonts and immature gametocytes of this species are very rarely seen in the circulating blood, as they are concentrated within the capillaries and blood sinuses of internal organs such as the brain, liver, kidneys, spleen, bone-marrow and especially—in pregnant women—the placenta. The reason for this seems to be that infected erythrocytes become sticky, and tend to adhere to each other and to the walls of the smaller blood vessels (see p. 112). Thus blood films made from persons infected with *P. falciparum* usually contain only very young trophozoites ('ring forms') and mature gametocytes.

The erythrocytes containing asexual parasites, except for those inhabited by the very young ring forms, show, when stained, a few, relatively large dots or lines called Maurer's clefts, but no Schüffner's or Ziemann's dots. Electron microscopy has shown that these clefts are within the cytoplasm of the erythrocyte and are probably cast-off portions of the parasite's outer membrane (Rudzinska and Trager, 1968), and it may be that Schüffner's and Ziemann's dots have a similar origin.

Pathology of malaria in man

It has already been stated that only *P. falciparum* directly causes fatal disease in man. The three other species, if untreated, cause recurrent fevers which are very debilitating and may lower the patient's resistance to other infection, but they do not, of themselves, kill him.

The fevers, so characteristic of malaria, occur when the schizonts in the erythrocytes burst, setting free their merozoites and also liberating into the blood various excretory products (including the malarial pigment) and the remains of the red cell cytoplasm. It is assumed that some of these liberated substances (not the merozoites themselves) are toxic and so induce a high fever, which is followed by rigors (violent shivering). The exact mechanism responsible for these effects is not, however, fully understood. The development of the schizonts tends to be synchronous, so that they all burst at the same time. The time elapsing between successive fevers is therefore often the same as the duration of the erythrocytic schizogony cycle, i.e. 72 hours with *P. malariae* and 48 hours with the other three species. Hence, the fevers produced by these species are said to be 'quartan' or 'tertian', respectively (see footnote on p. 105). Frequently during the first few days of an attack schizogony is less synchronized, and fevers may occur daily; but usually they soon adopt the rhythm characteristic of the species.

Exoerythrocytic schizonts are entirely non-pathogenic.[1] Thus, during the early part of the infection, before the erythrocytes are invaded, no symptoms develop. This is called the pre-patent period (the period before parasites can be detected in the blood). The period between infection and the appearance of the first *symptoms* is the incubation period. It clearly cannot be shorter than the pre-patent period, and must be longer by the duration of at least one erythrocytic schizogony cycle (since symptoms do not develop until the first erythrocytic schizonts burst); if the infection is slight, the incubation period may be still longer, since the presence of only a few rupturing schizonts in the blood may not produce noticeable symptoms.

If the infection is not treated and if death does not result rapidly (as it may with *P. falciparum*), after a variable number of erythrocytic schizogony cycles the host's antibodies will destroy most or all of the erythrocytic schizonts (exoerythrocytic schizonts are not affected by antibodies), all symptoms will disappear and eventually parasites will no longer be detectable in the peripheral blood: the infection is now latent. However, cure will not be complete: either a few erythrocytic schizonts survive in the capillaries of various viscera, or exoerythrocytic schizonts persist in some way in the liver (see p. 102), or both (except with *P. falciparum*, which has no persistent exoerythrocytic schizonts). Eventually, when either the level of circulating antibody has dropped or when the parasite itself has changed its antigenic structure, possibly by mutation, so that it is no longer affected by the existing antibody (Brown and Brown, 1966), the number of parasites in erythrocytes increases and a second clinical attack develops (compare the behaviour of *T. brucei* sspp., described in Chapter 3). In the absence of treatment, this process may continue for many years. Persons living in areas where malaria is common are regularly re-infected throughout their lives by the bite of infected mosquitoes and may, if they survive, develop incomplete resistance to, or partial tolerance of, the parasites; however, the infant death rate in such areas is appallingly high unless medical facilities are adequate to cope with rapid diagnosis and treatment.

Apart from the fever, malarial infection (in man and other animals) always leads to a massive increase in the number of phagocytic cells of the lymphoid-macrophage system. The spleen is the largest

[1] This is not true of some of the species of *Plasmodium* which infect birds.

agglomeration of these cells, and so it becomes grossly enlarged in chronic malaria. The spleen seems to be of great importance in the body's defence against malaria. In experimentally infected animals, in which the infection has become latent, surgical removal of the spleen leads to the rapid re-appearance of parasites in the blood.

P. falciparum exerts its lethal effect mainly by causing blockage of the capillaries in the viscera. The precise mechanism of this effect is not fully understood, but it seems to be due to the secretion by the parasites of a substance which constricts the capillaries, coupled with the fact that the surface membrane of infected erythrocytes becomes sticky, so that they tend to adhere to the capillary wall and to each other (see Maegraith, 1948 and Garnham, 1966c). This leads first to interference with the tissues' oxygen supply, and eventually to rupture of the blocked capillaries and bleeding into the surrounding tissue. This may occur in any or all of the internal organs, but is most serious in the brain, and it is damage to this organ that causes the death of most persons who die from acute malignant tertian malaria ('cerebral malaria'). Another very serious complication of *P. falciparum* infection is blackwater fever. This was associated with inadequate treatment with quinine, and is now fortunately rare. It consisted of wholesale destruction (lysis) of the patient's erythrocytes (for reasons incompletely understood), with excretion of the liberated haemoglobin in the urine (hence the term blackwater).

Diagnosis of malaria in man

Apart from the clinical findings (recurrent fevers, enlarged spleen), diagnosis depends on the demonstration of parasites in thick or thin blood films. There are no suitable experimental animals into which blood can be injected to detect scanty infections, nor any commonly used serological tests (though fluorescent antibody tests are useful in doubtful cases; see Voller, 1964).

Treatment and prevention of malaria in man

The drug quinine, an extract of the bark of the cinchona tree, was known for centuries in Peru before being imported into Europe during the first half of the seventeenth century. It is very effective in rapidly destroying the erythrocytic parasites, and is still often used in cases of cerebral malaria or where parasites have become resistant to other drugs. Because of the association of quinine with blackwater fever, however, general treatment is now usually given by synthetic drugs

Malaria parasites and their relatives

such as amodiaquine, pyrimethamine or chloroquine. Chloroquine, and another drug called proguanil, taken regularly in lower dosage, are chemosuppressants, i.e. they prevent clinical attacks of malaria. All these drugs are effective only against erythrocytic parasites, not the exoerythrocytic forms. Thus they do not radically cure infections due to *P. vivax, P. malariae* or *P. ovale.* This is best attempted by primaquine, a rather more toxic drug which should be used only in hospitals.

Apart from chemoprophylaxis, malaria can, of course, be prevented by avoiding contact with infective mosquitoes. Since all the vector species of *Anopheles* feed in the evening or at night, this can be done by screening the windows and doors of houses with fine-mesh netting and by the use of mosquito-nets over beds. Also, if adequate funds are available, the mosquitoes can be destroyed, and malaria eradicated, by spraying houses with long-lasting insecticides and by draining the swamps, etc. where they breed. Thanks to the efforts of local health authorities, governments and the World Health Organization, much progress has been made in eliminating malaria from towns, districts and even whole countries where it was previously a scourge; but much still remains to be done.

OTHER SPECIES OF *Plasmodium*

Some of the many species of *Plasmodium* which infect animals other than man are used for experimental work. The more important of these are listed below.

(1) *Plasmodium (Plasmodium) cynomolgi.* This species infects monkeys (mainly *Macaca* spp.) in India, Ceylon and the Far East. It is closely similar to *P. vivax,* and several subspecies exist, at least some of which can also infect man.

(2) *Plasmodium (Plasmodium) knowlesi.* Another parasite of (mainly) *Macaca* spp. in Asia, this species is unique among those infecting primates in having a 24-hour erythrocytic schizogony cycle. It is lethal to rhesus monkeys, and can infect man (in whom it causes only mild disease).

Several other species have been described from Asian and South American monkeys. In Africa only one species is known in monkeys —*P. (P.) gonderi*—but others infect the higher apes (chimpanzees and gorilla). Usually they are not markedly pathogenic.

(3) *Plasmodium (Vinckeia) berghei.* One of the only two known species of malaria parasitic in murine rodents, *P. berghei* was discovered in the Congo. It can be maintained in laboratory rats and mice, and is fairly pathogenic to them.

(4) *Plasmodium (Haemamoeba) gallinaceum* (Figs. 108, 109). A natural parasite of the jungle-fowl of Asia, this parasite is commonly maintained in chickens in laboratories and is often used

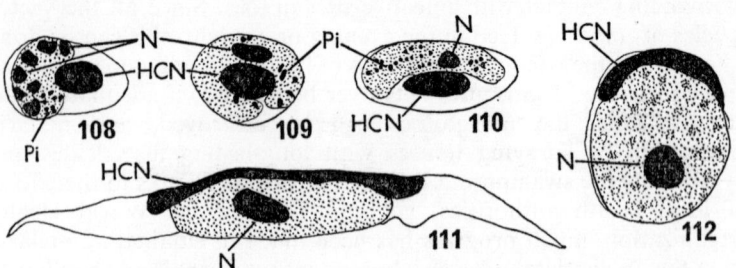

Fig. 108. *P. gallinaceum*; schizont in chicken's erythrocyte. ×1,500.

109. *P. gallinaceum*; macrogametocyte (female) in chicken's erythrocyte. ×1,500.

110. *Haemoproteus palumbis*; macrogametocyte in erythrocyte of English wood-pigeon. ×1,500.

111. *Leucocytozoon simondi*; elongated macrogametocyte in blood-cell of Canadian duck. ×1,500.

112. *L. simondi*; round macrogametocyte. ×1,500.

Abbreviations: HCN – host cell's nucleus; N – nucleus (of parasite); P – parasite; Pi – pigment.

in testing possible new antimalarial drugs. It can cause outbreaks of disease in flocks of domestic hens and often kills the younger birds, as a result of blockage of the brain capillaries by the large secondary exoerythrocytic schizonts, which develop in the endothelial cells.

The majority of the species of *Plasmodium* which infect birds and reptiles do not seem to be very pathogenic.

Family Haemoproteidae

Members of this family are very similar to *Plasmodium* in their morphology and life cycles, but differ in not having erythrocytic schizogony and in being transmitted by insects other than mosquitoes. The gametocytes, which are the only stages to be found in erythrocytes, contain pigment and cannot be distinguished at the generic level from the gametocytes of *Plasmodium*. Exoerythrocytic schizonts of members of this family are found in various tissues, and are sometimes quite large. Haemoproteids have been described from mammals (mainly primates), birds and reptiles. The majority of species are only slightly, if at all, harmful to their hosts. Three of the more important genera are listed below.

Hepatocystis Levaditi and Schoen, 1932
This is the commonest haemoproteid of mammals. The exoerythrocytic schizonts, in liver parenchyma cells, are very large (up to 1 mm) and are called merocysts. The life cycle of only one species (*H. kochi* of African monkeys) is known; it is transmitted by *Culicoides* spp. (midges: Diptera, family Ceratopogonidae).

Haemoproteus Kruse, 1890
A common parasite of birds, throughout the world. Several species have been described. *H. palumbis* (Fig. 110) occurs frequently in English wood pigeons *(Columba palumbus)*. *Haemoproteus* is transmitted by large ectoparasitic flies of the family Hippoboscidae ('louse-flies': Diptera), and schizogony occurs chiefly in the lung.

Parahaemoproteus Bennett, Garnham and Fallis, 1965
Another common genus found in birds, including ducks in North America. Schizogony is mainly in viscera other than the lung, and the vectors, where known, are midges *(Culicoides)*.

Family Leucocytozoidae
Members of this family are known only from birds, with one exception described recently from a reptile by Lainson and Shaw (1969). As in the Haemoproteidae, only the gametocytes are found in blood cells. Originally they were thought to inhabit leucocytes, but now they are known, in at least some species, to infect the precursors of the erythrocytes. The infected cells become enlarged, and their nuclei are considerably altered, so that only while the contained parasite is very young is it possible to identify the host cell. Infection with certain species results in a curious but characteristic elongation of some of the host cells, which become spindle-shaped; the contained gametocyte is elongated (Fig. 111). Other gametocytes (probably produced by a different type of schizont) and their host cells are rounded (Fig. 112). The gametocytes of all species in this family are larger than those of the Haemoproteidae and Plasmodiidae, and differ in not having any malarial pigment; as it is now known

that they sometimes inhabit erythrocytes, this must reflect a difference in their metabolism. The life cycle of leucocytozoids is basically similar to that of the two other families in the Haemosporina. Exoerythrocytic schizonts are usually large, and may occur in various internal organs. Three genera are recognized.

Leucocytozoon Ziemann, 1898

Until recently, this was thought to be the only genus in the family. Many species have been described, from birds the world over. Most are seemingly harmless, but one, *L. simondi* (Figs. 111, 112), causes severe disease and death in domestic ducks in North America while wild ducks serve as a reservoir of infection and, as is so often true, seem to be less severely affected. Many wild birds in Britain are infected with various species. Where known (and they are unknown for all the British species[1]), the vectors are blackflies of the genus *Simulium* (Diptera, Nematocera, Simuliidae).

Akiba Bennett, Garnham and Fallis, 1965

This genus contains a single species, *A. caulleryi*, which causes disease in chickens in Japan and eastern Asia. It was previously regarded as a species of *Leucocytozoon*, but was re-classified largely on the grounds of its transmission by *Culicoides* instead of *Simulium;* whether this separation was justified remains to be determined.

Saurocytozoon Lainson and Shaw, 1969

This genus, with a single species, was described from a Brazilian lizard by Lainson and Shaw (1969). Morphologically it resembles a species of *Leucocytozoon* with round gametocytes.

---

[1] Recently a species of *Simulium* has been shown to transmit *L. Sakharoffi* of English rooks.

8

PIROPLASMS

These organisms can be simply defined as parasitic Protozoa inhabiting the erythrocytes and sometimes other cells of vertebrates, which do not form pigment from the erythrocyte haemoglobin. As far as is known, they are all transmitted by ticks (Arthropoda, Acarina, families Ixodidae and Argasidae), though other vectors may remain to be discovered, particularly for those piroplasms whose hosts are aquatic. All the piroplasms are small, usually round or pear-shaped when in their host's erythrocytes (hence the name piroplasm). They are known from fish, amphibia, birds and mammals. None of them naturally infects man, though occasional accidental infections have been recorded (see p. 120). Their correct taxonomic position has been much debated. For many years they were classified in the Sporozoa, very close to the malaria parasites (see, e.g. Wenyon, 1926). When their life cycle in the tick vectors began to become known, considerable differences became apparent between it and the sporogony of malaria, and taxonomists tended to remove the piroplasms from the Sporozoa and leave them in limbo (e.g. Levine, 1961). The next step was to assign the group to the Sarcodina, on grounds which to the writer, at least, were never convincing (Honigberg *et al.*, 1964). Finally, studies on their ultrastructure are now leading back to the view that the piroplasms are indeed Sporozoa, though perhaps less closely related to the Haemosporina than was previously thought (Büttner, 1967; Friedhoff and Scholtyseck, 1968).

Following this latter trend, the piroplasms will here be treated as a separate class, Piroplasmea, of the subphylum Sporozoa. Within the class, all known genera are grouped in a single order **Piroplasmida**, which is divided (by many but not all authorities) into three families —Babesiidae, Theileriidae and Dactylosomidae.

Family Babesiidae

These are piroplasms parasitizing reptiles, birds and (mainly) mammals, in which they inhabit only erythrocytes. Multiplication is by binary fission, or by schizogony resulting in the formation of four merozoites. Opinions differ widely (and wildly!) as to the number of genera amongst which the species in this family should be shared, ranging from one to fourteen genera. A compromise solution of two is adopted here, *Babesia* and *Echinozoon*. The latter is a rare parasite of the rock hyrax in Sudan, and is characterized by the fact that infected erythrocytes develop about twenty long filamentous extrusions from their surface.

Genus *Babesia* Starcovici, 1893

Some of the commoner species of *Babesia* of veterinary importance, their main hosts and geographical distribution, are listed in Table 9; two are illustrated in Figs. 113 and 114. In addition, many species have been described from wild rodents (including those in England; see Shortt and Blackie, 1965) and other animals. Many of these

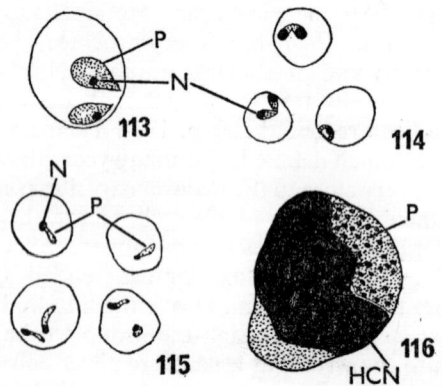

Fig. 113. *Babesia canis*; two parasites in erythrocyte of dog. × 1,500.

114. *B. divergens*; parasites (two undergoing division) in erythrocytes of ox. ×1,500.

115. *Theileria parva* in erythrocytes of ox. ×1,500.

116. *T. parva*; schizont in lymphocyte of ox. ×1,500.

Abbreviations: HCN – host cell's nucleus; N – nucleus (of parasite); P – parasite.

Species	Size[1]	Geographical distribution	Main hosts Vertebrate	Main hosts Invertebrate	Pathogenicity	Common name of disease (if any)
Babesia bigemina	Large	Central and S. America, Europe, Africa, Australia	Cattle, deer	*Boophilus*, *Haemaphysalis*, *Rhipicephalus*	High	Red-water fever
B. bovis	Small	Europe, Russia, Africa	Cattle, deer	*Ixodes*, *Boophilus*, *Rhipicephalus*	Moderate	Red-water fever
B. divergens	Small	W. and Central Europe (including England)	Cattle	*Ixodes*	Moderate	Red-water fever
B. argentina	Small	S. and Central America, Australia	Cattle	*Boophilus*	High	
B. major	Large	Europe, Russia	Cattle	*Boophilus*	Low	
B. caballi	Large	S. Europe, Asia, Russia, Africa,	Equidae (domestic)	*Dermacentor*, *Hyalomma*, *Rhipicephalus*	Moderate	
B. equi	Small	S. Europe, Asia, Russia, Africa, S. America	Equidae (domestic and zebra)	*Dermacentor*, *Hyalomma*, *Rhipicephalus*	High	Biliary fever
B. motasi	Large	S. Europe, Russia, Africa, Far East, tropical America	Sheep, goats	*Rhipicephalus*, *Haemaphysalis*, *Dermacentor*	Moderate	
B. ovis	Small	As for *B. motasi*	Sheep, goats	*Rhipicephalus*, *Ixodes*	Low	
B. trautmanni	Large	S. Europe, Africa, Russia	Pig	*Rhipicephalus*	Moderate	
B. canis	Large	N., Central, S. America, S. Europe, Russia, Africa, Asia	Dogs and wild canids	*Rhipicephalus*, *Dermacentor*, *Haemaphysalis*	Moderate-high	Tick fever
B. gibsoni	Small	India, Ceylon, China	Dogs and wild canids	*Rhipicephalus*, *Haemaphysalis*	High (in domestic dog)	Tick fever
B. felis	Small	Africa, India	Domestic cat, lion, leopard	? *Haemaphysalis*	Moderate	

[1] In this Table, 'large' means about 2–5 μ long, and 'small' means about 1–2 μ.

species are quite benign, though others produce disease (babesiasis) in domestic animals (particularly cattle and dogs). *B. rodhaini,* a natural parasite of wild rodents in Africa, is often maintained in laboratory mice for use in experimental work.

Human babesiasis

Although not normally parasitic in man, three human infections with babesiids have been recorded. The first of these was in 1956, in a Jugoslav farmer, whose spleen had been removed after an accident eleven years earlier: he developed a fatal infection with *Babesia bovis* (or *B. divergens*). The second case was an American hunter, who in 1966 became infected with an unidentified babesiid. He was at first thought to have malaria, and so was treated with chloroquin and promptly recovered. A third, fatal, infection was recently reported from Ireland (Fitzpatrick *et al.,* 1968). Both these men, also, had had their spleens removed some time previously. Thus it appears that the spleen is the organ responsible for preventing infection of man with babesiids. This hypothesis was tested by removing the spleens of two chimpanzees and one rhesus monkey and then inoculating them with two species of *Babesia* to which they are normally entirely resistant: all the animals became infected (Garnham and Bray, 1959; Shortt and Blackie, 1965).

Morphology and life cycle

Morphological differences between different species of *Babesia* are slight. They all appear in erythrocytes as round or oval organisms, ranging in size from 1 to 5 or 6 μ in maximum diameter. Division stages are often seen (Fig. 114) forming either 2 or 4 merozoites; species dividing into four are often grouped into a separate genus, *Nuttallia,* but this is apparently an artificial distinction since both types of division may occur together (see Shortt and Blackie, 1965). Their life cycle in the tick vector is very complicated and, even now, not fully understood. (This is in spite of the fact that a species of *Babesia* was the first parasite which was shown to be transmitted by an arthropod, by Smith and Kilborne in 1893; see Riek, 1964.) Some species of tick feed only once during each stage (larva, nymph, adult) of their life cycle and in these species, the piroplasms are passed from stage to stage so that such a tick which, for example, becomes infected as a larva, will transmit the infection only after moulting and becoming a nymph. Some babesiids may also enter the eggs of adult female ticks and so pass to the larvae before being transmitted to another vertebrate. No clear evidence has yet been obtained of the existence of any sexual process in the life cycle of piroplasms, though its occurrence is suspected by many workers. The most recent, and one of the fullest, accounts of the life cycle of a babesiid is that of Riek (1964) for *B. bigemina*. After being ingested in the blood on which an adult female tick is feeding, the parasites enter the cells lining the tick's gut and undergo multiple fission to produce motile stages resembling the ookinetes of *Plasmodium* spp. (but lacking pigment). Some of these migrate into cells in the haemolymph and thence to the cells of the Malpighian tubes, where they again divide. Eventually the motile forms enter the eggs. When the larva hatches, they continue dividing in its gut cells. After the larva moults and becomes a nymph, the parasites migrate to the salivary glands, enter their cells and again divide by multiple fission to produce, 8–10 days later, large numbers of small ($2 \cdot 5 \times 1 \cdot 25$ μ) infective forms, which enter the mammalian host in the nymph's saliva.

Pathogenesis

When species of *Babesia* are pathogenic, the disease is usually associated with anaemia, fever, enlargement of the spleen (which is not dark in colour, as it is in malaria, because of the absence of malarial pigment in babesiasis), and blocking of the capillaries in various tissues (including the brain), which may damage the cells by depleting their oxygen supply (as in malaria due to *P. falciparum*). The anaemia is often accompanied (again, as in malignant tertian malaria) by the lysis of erythrocytes and excretion of the released haemoglobin in the urine; hence, babesiasis of cattle is known as red-water fever. Animals which recover from an acute attack are usually premune (i.e. resistant to re-infection so long as they remain chronically infected, which may be for two years—or more, if they are constantly being bitten by infected ticks); relapses are not known to occur.

Diagnosis

Infections are diagnosed, apart from the clinical signs and symptoms, if any, by finding parasites in blood films. They may be difficult to distinguish from 'ring' forms of *Plasmodium* species, especially in thick films, but if division forms are seen this confusion should not occur as malarial schizonts are almost always considerably larger, and produce more daughter individuals, than dividing piroplasms; they also contain pigment.

Treatment and prevention

A variety of drugs is used in treating babesiasis of domestic animals, including some of those active against trypanosomes (e.g. 'Berenil', 4, 4'-diamidino diazoaminobenzene diaceturate). Others used include trypan blue (not effective against all species) and acriflavine.

Babesiasis of domestic animals can best be prevented by the use of insecticides to keep the animals free from ticks. Further details are given by Richardson and Kendall (1963).

Family Theileriidae

The forms inhabiting erythrocytes are very small (1–2 μ), and are similar to those of *Babesia,* but members of this family are separated from the babesiids by the fact that they also inhabit other cells (usually lymphocytes), in which they undergo schizogony. The intraerythrocytic forms of some, if not all, species divide into two or four, as in *Babesia.* The type and location of exoerythrocytic schizogony is used as the basis for dividing the family up into two genera, as follows.

(1) *Theileria.* The exoerythrocytic schizonts are in lymphocytes, and are small or medium-sized (about 10–20 μ).

(2) *Cytauxzoon.* The exoerythrocytic schizonts are in fixed tissue cells, and are large (over 50 μ when mature) but divided into cytomeres.

The accuracy of this simple classification is in some doubt; probably as more becomes known about the Theileriidae, their taxonomy will have to be revised. The theileriids seem to parasitize ruminants alone (except of course for their tick vectors), mainly but not exclusively in the tropics. Only *Theileria* infects domestic animals, *Cytauxzoon* spp. having been recorded so far from only three species of African wild ungulates (duiker, kudu and eland).

Genus *Theileria* Bettencourt, França and Borges, 1907[1]

The important species of this genus are listed in Table 10, together with their main hosts and geographical distribution. Morphologically the erythrocytic forms (Fig. 115) resemble small babesiids, about

[1] *Gonderia* is here regarded as a synonym of *Theileria.*

TABLE 10
THEILERIIDAE OF VETERINARY IMPORTANCE

Species	Geographical distribution	Main hosts		Pathogenicity	Common name of disease (if any)
		Vertebrate	Invertebrate		
Theileria parva	East Africa (eradicated from southern Africa)	Cattle, buffalo	*Rhipicephalus*, *Hyalomma*	High	East coast fever
T. annulata	N. Africa, S. Europe, S. Russia, India, China	Cattle, water buffalo	*Hyalomma*	High	Mediterranean coast fever
T. mutans	Africa, Asia, Europe (including England), Russia, Australia, N. America	Cattle	*Rhipicephalus*, *Boophilus*	Very low	
T. hirci	N. Africa, S.E. Europe, S. Russia, Middle East	Sheep, goat	? *Rhipicephalus*	High	
T. ovis	Africa, Europe, Russia, India, Middle East	Sheep, goat	*Rhipicephalus*	Very low	

1 to 2 μ long by 0·5–1 μ wide. It is not always easy to distinguish them, in a blood film, from some of the smaller babesiids.

The exoerythrocytic schizonts are found chiefly in the spleen and lymph nodes, though they may also be seen wherever there are lymphocytes (i.e. in all viscera). They are irregular in shape, usually round or oval, and measure about 10–20 μ in diameter when fully grown (Fig. 116). They contain a mass of cytoplasm[1] with a number of nuclei, and finally split into many small uninucleate bodies (merozoites) which invade the erythrocytes. When an infected lymphocyte divides, both daughter cells appear to retain part of the schizont and so the number of infected lymphocytes increases.

The vectors, in all known cases, are ixodid ticks. The development undergone in the tick is even less well known than is that of the babesiids. The infection can pass from stage to stage of the tick, but not (as far as is known) through the tick's eggs. In *T. parva*, many of the ingested parasites probably die but some enter gut epithelial cells (probably the cells ingest them phagocytically) and in a recent study (Martin *et al.*, 1964), development was observed here. Somehow the parasites enter the secretory cells of the tick's salivary glands, perhaps after the moult to the next stage (i.e. nymph or adult), and here they multiply (possibly by multiple fission or a form of schizogony). The progeny of this multiplication are infective to susceptible mammals, which they enter in the saliva of the tick when it next feeds. The duration of this cycle is not known, but is said to be short. No firm evidence for the existence of sexual processes in this developmental cycle has yet been obtained.

Pathogenesis

The theileriases occurring in domestic cattle, sheep and goats vary from mild to acute, often fatal, febrile diseases with enlargement of the lymph nodes and spleen, and congestion of the lungs and meninges (the membranes surrounding the brain). Ulcers sometimes develop in the abomasum and intestine. Anaemia may occur if the infection is heavy. Haemoglobinuria is much less common than in babesiasis though it may occur in disease in cattle due to *T. annulata* and, briefly, in that due to *T. hirci* of sheep and goats. The most pathogenic species are *T. parva* (mortality rate up to 100%) and *T. annulata* in cattle, as well as *T. hirci* in sheep and goats.

---

[1] The bright blue colour of this cytoplasm after Giemsa's staining has led to their being colloquially called 'Koch's blue bodies'.

The species found in English cattle (*T. mutans*) is only slightly pathogenic, if at all.

Diagnosis

Diagnosis of all these species depends on the finding of parasites in blood films (when the remarks about *Babesia* on pp. 121–122 above also apply) and also in smears of material obtained by lymph glan dpuncture where exoerythrocytic schizonts may be seen.

Treatment

No effective treatment is known for any of the theileriases, so outbreaks due to one of the more pathogenic species may have devastating results. Prevention can only be by attacking the tick vector by dipping and spraying the animals, good fencing to keep out wild ungulates which may serve as reservoirs of infection (e.g., buffalo, in the case of *T. parva* in East Africa and quarantine measures directed against outbreaks (including sometimes the slaughter of infected herds). Animals recovering from infection are resistant to re-infection, though with all species except *T. parva* this appears to be dependent upon the persistence of small numbers of viable parasites in the host, and hence such animals are potential reservoirs of infection (Levine, 1961). Vaccination with live, avirulent strains has been used with some success to prevent *T. annulata* infection.

Family Dactylosomidae

This family contains piroplasms of cold-blooded vertebrates, of which two genera have been described. Little is known of the pathogenicity and life cycle of these genera. They do not appear to harm their hosts, and only stages in erythrocytes have been seen. Nothing is known of their transmission. Schizonts are seen in the red blood cells, and also non-dividing forms which are thought (with no direct supporting evidence) to be gametocytes. The two genera which have been described are *Dactylosoma* Labbé, 1894 (recorded from reptiles, amphibia and fish in various, scattered parts of all the continents except Australasia) and *Babesiosoma* Jakowska and Nigrelli, 1956 (so far recorded only from amphibia in North America and fish in Africa). *Dactylosoma* produces from four to sixteen merozoites, arranged in a fan-like pattern within the erythrocytes, while *Babesiosoma* has schizonts closely resembling those of some species of *Babesia,* with only four merozoites. For a fuller account of these two genera, see Jakowska and Nigrelli (1956).

Several other 'genera' which were at one time thought to belong to this class, or close to it, are now known to be not Protozoa but bacteria *(Bartonella),* rickettsiae *(Anaplasma, Eperythrozoon, Haemobartonella* and *Aegyptianella)* or viruses *(Pirhemocyton);* see Levine (1961), Stebhens and Johnston (1966) and Bird and Garnham (1967).

9

TOXOPLASMEA

The organisms composing this class have been shuffled around by taxonomists between all the protozoan groups (and even the fungi) at various times, but more detailed knowledge of their life histories and their micro-anatomy (as revealed by the electron microscope) has resulted in general agreement that they are best placed in the subphylum Sporozoa as a separate class, the Toxoplasmea. Within this grouping, the three genera which compose the class are usually regarded as members of a single order, the Toxoplasmida. Each genus may be relegated to one of the three families Toxoplasmatidae, Besnoitiidae and Sarcocystidae, as suggested by Garnham (1966a). Other workers, however, unite the members of the first two families as Toxoplasmatidae (Levine, 1961). The former (separatist) view is followed here.

Since the Toxoplasmatidae contains the only important pathogen of man in the class, study has been concentrated on this family. The life cycles of the members of the other two families are very imperfectly known.

The class Toxoplasmea may be defined as Sporozoa lacking resistant or transmissive spores and possessing cysts or pseudocysts (see below) containing large numbers of individuals (zoites). The single order, Toxoplasmida, has the same characters.

Family Toxoplasmatidae
Genus *Toxoplasma* Nicolle and Manceaux, 1909
Several species have been described in this genus, including *T. hominis* from man, but (with one exception) all are now believed to be

synonyms of the earliest, *T. gondii*. The exception is *T. microti*, a parasite of wild voles, the true systematic position of which is still in doubt. It is probably more nearly related to *Sarcocystis* (see p. 133).

Toxoplasma gondii. First described from a wild rodent *(Ctenodactylus gundi)* in North Africa in 1908, this parasite has since been recorded from a variety of mammalian and avian hosts. Different names have been given to various strains, but all evidence supports the view that they represent a single species. The parasite can be freely transmitted (by experimental inoculation) from one host species to another; there is no morphological difference between the organisms found in different host species; and there is no consistent biological difference. The various strains may differ in their virulence (i.e., their ability to increase in number in the host), but this is labile and can sometimes be changed by passage through a different host (see Beattie, 1964).

T. gondii thus has an unusually wide host-range for a protozoan parasite: it can probably infect all warm-blooded animals (mammals and birds), but not cold-blooded ones. This parasite is equally widespread geographically and has been recorded from all parts of the world. It may cause acute illness (toxoplasmosis) and death of infected persons (or other animals), or the infection may be, and indeed usually is, latent and completely inapparent. It is likely that about one in every four persons reading this book has been infected with *T. gondii* at some time, without having been aware of it (see p. 132).

Fig. 117. *Toxoplasma gondii*; zoites liberated from host cell in a smear of mouse peritoneal fluid. ×1,500.

118. *T. gondii*; pseudocyst within macrophage in smear of mouse peritoneal fluid. ×1,500. (Note individual at arrow, apparently undergoing endodyogeny.)

119. *T. gondii*; cyst in smear of brain of mouse (slightly diagrammatic, not all zoites shown). ×750.

120. *Besnoitia besnoiti*; zoites from pseudocyst in section of skin of ox. ×1,500.

121. *B. besnoiti*; pseudocyst in section of skin of ox. ×150.

122. *Sarcocystis lindemanni*; cyst in section of human leg muscle (Dr Abbott's case). ×150.

123. *Sarcocystis* sp.; zoites liberated from a cyst. ×1,500.

124. *Pneumocystis carinii*; cyst in smear from lung of cortisone-treated rat. ×1,500. (Slide kindly loaned by Mr R. Killick-Kendrick.)

Abbreviations: Cy – cyst; Ep – epidermis; De – dermis; HCN – host cell's nucleus; N – nucleus; P – parasite; Pc – pseudocyst; Sn – region of sarconemes (anterior end).

Toxoplasmea

Morphology and life cycle

Only one stage is known (but see p. 131), a small crescentic organism called a zoite (sometimes 'trophozoite'), slightly pointed at one end with a central nucleus (Fig. 117). The zoite is about $5\,\mu$ long and $1-2\,\mu$ broad, and it can move with a gliding motion, the mechanism of which is not

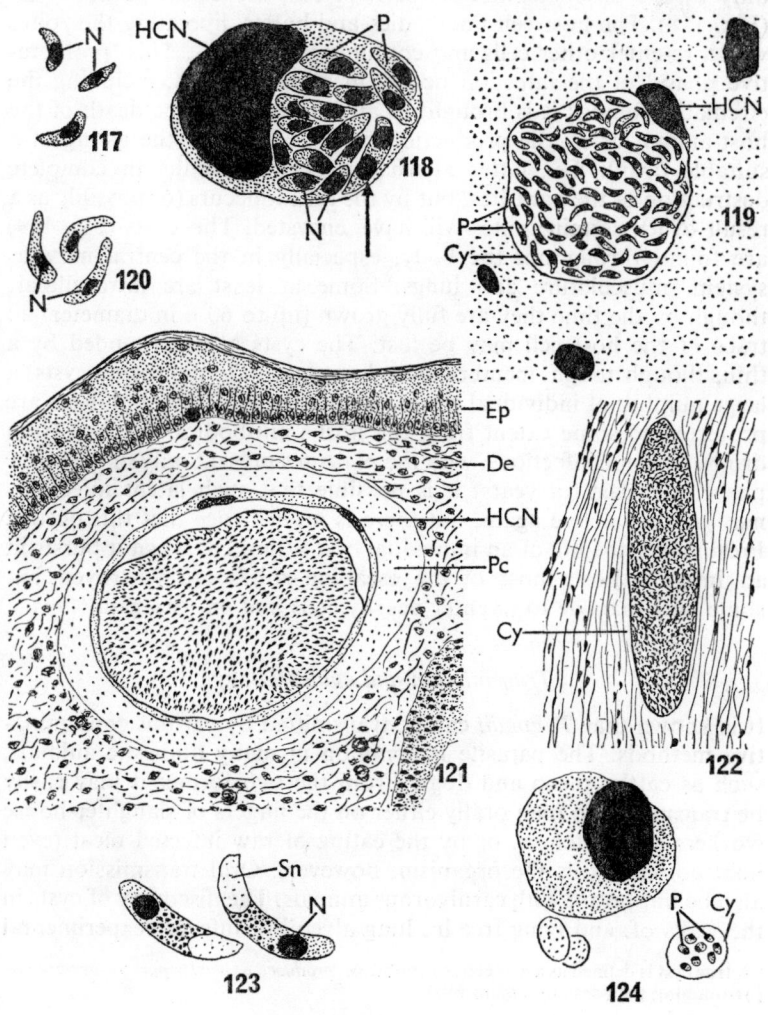

known, and it can also bend its narrower (anterior) end. Accounts of its reproduction vary, but it appears to divide by a process called endodyogeny, or internal budding, in which two 'daughters' develop within the parent. There is no evidence of any sexual process.

In the early stage of infection (in any animal), the zoites enter macrophages (both actively and by ingestion) and divide until the cell is full of them. These aggregations of parasites, bounded only by the plasmalemma of the host cell, are called pseudocysts[1] (Fig. 118). The host cell finally dies and bursts, liberating the zoites which re-enter other cells and continue the process. This 'proliferative phase' of the infection occurs in all the viscera, including the circulating blood; it is brought to an end either by the death of the host-animal or by the production of antibodies (or the giving of a suitable drug). The latter alternative seldom results in complete destruction of the parasites, but by the time it occurs (or possibly as a result of it), certain zoites will have encysted. The cysts (Fig. 119) are found throughout the body, especially in the central nervous system, musculature and lungs. Some at least are intracellular, though by the time they are fully grown (up to 60 μ in diameter) all trace of the host cell may be lost. The cysts are surrounded by a thin, though tough, membrane and contain (like the pseudocysts) a large number of individual zoites. Within the cyst the organisms are protected to some extent from the host's antibodies and also from drugs. Chronic infections, with the parasites present only in the cystic phase, may last for years; it is possible that such latent infections may become active again, but there is no evidence that this occurs. Presumably the life of an individual cyst is limited; when it bursts in an immune host, most of the escaping zoites are destroyed but some may manage to re-enter nearby cells and encyst again.

Transmission and epidemiology

It is known that *T. gondii* can be transmitted by various contaminative methods. The parasite is common in many domestic animals, such as cattle, sheep and dogs. From the two former it may easily be transmitted to man orally either on the fingers of slaughter-house workers and butchers, or by the eating of raw infected meat (even light cooking kills the organism, however). Oral transmission may also be important with carnivorous animals. The discovery of cysts in the walls of, and lying free in, lung alveoli of infected experimental

---

[1] A true cyst is defined as a protective membrane, *produced at least in part by the parasite*, surrounding the parasitic organism(s).

animals suggests that droplet infection, and infection from nasal mucus and saliva, may be common between man and man, dog and man, and other animals also. A well-known means of infection, in man and other animals, is the congenital route. If a pregnant female has an acute infection, the chance of the parasites passing through the placenta and infecting the embryo is fairly high though how this occurs is not known. Recently, a considerable amount of evidence has been produced suggesting that trans-placental infection can also occur if the mother's infection is latent, but this must be less common. Parasites have been isolated from the uterine wall and also from the placenta itself in both human beings and experimental animals (see review by Jacobs, 1967).

Recently, Hutchison and his colleagues (Hutchison *et al.*, 1968; Work and Hutchison, 1969) have shown that *T. gondii* can be transmitted in the faeces of infected animals, where an encysted stage of the parasite has been identified. This method of transmission is probably responsible for the majority of post-natal infections, particularly in herbivorous animals.

Pathogenesis

In man and, as far as is known, in other animals, toxoplasmosis which has been acquired after birth results in fever and swelling of the lymph glands, and may be so mild as to pass unnoticed. The disease may progress to a more generalized infection, in which brain, lungs, liver and other tissues are involved, and death may result from the damage caused to these organs (particularly the brain). This is the acute phase of toxoplasmosis. In experimentally infected mice, one of its characteristics is a pronounced increase in the amount of peritoneal fluid, which contains large numbers of infected macrophages (pseudocysts). If death does not ensue, the acute infection proceeds to the chronic, latent phase, in which only encysted parasites are present and no sign of disease is seen. Since there is serological evidence (see p. 132) that about one quarter to one third of the population of this country has at some time been infected with *T. gondii*, yet fewer than 200 clinical cases occur annually (Beattie, 1964), it is obvious that the majority of infections are indeed very mild, and the development of severe or fatal illness a very rare consequence.

This is unfortunately not true of infections acquired before birth, though these themselves are happily quite rare (probably less than one per 1,000 live births). By the time the infected child is born, infection may be almost entirely restricted to its brain, where it often devastates large areas of tissue, possibly due to an allergic reaction to its presence or to liberation of a (hypothetical) toxin; the retina of the eye is characteristically infected in such cases, and this may lead to blindness. In relatively mild congenital toxoplasmosis, damage to the retina may be the only detectable sign. Severely afflicted infants are often either still-born, or die soon after birth. A congenitally infected child may be born with severe generalized infection accompanied by jaundice resulting from damage to the liver. It is believed that this results from infection having occurred later in pregnancy: the child with infection restricted to the brain and eye was presumably infected earlier and the initial generalized infection was over, by the time of birth.

Diagnosis

A clinical diagnosis of suspected toxoplasmosis is best confirmed by the isolation of the parasite from material such as lymph gland or tonsil tissue (in man) or any viscera (in other animals), by inoculation into mice. There are various serological procedures which detect the presence of antibodies to *T. gondii;* these do not necessarily indicate the presence of infection (even latent) at the time of making the test, since some antibodies probably persist for some years (at least) after cure, but a positive serological test indicates experience of the parasite at some time. If two serological tests are done at intervals of a few weeks, and the second shows a higher titre (i.e. a more strongly positive reaction) than the first, then acute toxoplasmosis is probable. The most commonly used tests include a complement-fixation reaction, using an extract of *T. gondii* zoites as antigen, an agglutination test (in which killed zoites are caused to adhere to one another when treated with serum containing antibody), and the so-called 'dye test' which depends on the fact that specific antiserum affects living zoites in such a way that they do *not* become stained when immersed in methylene blue solution, whereas normal living zoites are stained. References to fuller accounts of these and other tests are given by Beattie (1964). Such tests (the dye test in particular) have given positive results in up to 77% of cats, 32% of dogs, 50% of pigs, 64% of sheep, 21% of cattle, and 25–30% of human beings in England and the United States of America while higher figures have been obtained for man in other regions—e.g. 94% in Guatemala (figures derived from Beattie, 1964). *Toxoplasma* is indeed a common parasite but a rare pathogen.

Treatment

Severe congenital toxoplasmosis has usually done its damage before birth, so treatment is of little use. Post-natal infection, however, can be controlled (though not usually completely cured) by a mixture of one of the less-used anti-malarial drugs, pyrimethamine, with sulphonamides.

Family Besnoitiidae
Genus *Besnoitia* Henry, 1913

This genus is classified by Levine (1961) in the family Toxoplasmatidae, but the reasons for separating it put forward by Garnham (1966a) seem valid. Little is known of the genus. Different species infect cattle, horses, deer, rodents and lizards, but not man. All known species are seen as zoites (Fig. 120), like those of *T. gondii* but a little larger ($5-9 \times 2-4$ μ), inhabiting large (100–500 μ) pseudocysts (Fig. 121), which have a thick collagenous wall containing numerous host nuclei, probably belonging to an original single host cell (possibly of the lymphoid-macrophage series) which has become greatly enlarged and multinucleate; within the pseudocyst the zoites multiply by binary fission or, possibly, endodyogeny. Nothing is known of the life cycle or transmission of this genus, although *B. jellisoni* (of North American rodents) can be transmitted experimentally by feeding or injecting visceral material containing pseudocysts. *B. jellisoni* will infect (and may kill) laboratory mice, and the pseudocysts occur in the mesenteries.

B. besnoiti of cattle (in Europe and Africa) and *B. bennetti* of horses (Europe, Africa and North America) inhabit the skin (and the cornea). They are of some economic importance since, although the disease which they produce is chronic, infected animals lose condition and the hair drops out of infected areas of skin. The mortality rate is about 10% and no reliable treatment is known.

Family Sarcocystidae
Genus *Sarcocystis* Lankester, 1882

Many species of this genus have been described. It is very common (70–100%) in some herbivorous mammals (cattle, sheep, horses), and has also been recorded in pigs, monkeys, rabbits, rodents, ducks, chickens, man (rarely) and many other species throughout the world. Like *Besnoitia,* little is known of the biology of this genus. Almost always it seems entirely non-pathogenic, though fatal infections have been recorded in mice, and in man it may cause mild disease (p. 135). Its method of transmission is unknown. The experimental feeding of infected material, and of faeces from infected animals, has sometimes resulted in transmission, and Levine (1961) thinks that faecel transmission is the natural method. Although this has been accomplished only very rarely, recent findings with *Toxoplasma* (see p. 131) suggest that it may be a genuine means of transmission of *Sarcocystis*.

All species are almost entirely restricted to the muscle fibres (including cardiac muscle) of their hosts (rarely they have been recorded from brain). Here they are seen as large, sometimes very large, oval or elongated cysts (or pseudocysts?) which may be as long as 1–2 mm (Fig. 122). The larger cysts are divided into irregular compartments by a network of cytoplasmic partitions called trabeculae, and their central regions may be more or less empty. The cyst is lined by a layer of parasite cytoplasm which contains many nuclei, not separated from one another by cell walls. It is from this lining layer that the trabeculae develop. The cyst wall itself is complex: on the outside is a layer of host connective tissue, and within this a spongy or fibrous layer (or layers) of uncertain origin (hence the doubt as to whether this body should be called a cyst or pseudocyst; but if one regards the lining layer of cytoplasm as a part of the wall, the term 'cyst' would be correct, and will be used here). From the layer of nucleated cytoplasm lining the cyst wall are budded off rounded cells which divide and give rise to the zoites which divide by endodyogeny. The zoites (Fig. 123) closely resemble those of *Toxoplasma* and *Besnoitia* in their general shape and structure, but they are larger (10–15 μ long).

Various descriptions exist of alleged developmental stages of *Sarcocystis* in the intestinal mucosa (see Levine, 1961, and Wenyon, 1926), but these are incomplete and some at least may not be *Sarcocystis* at all.

S. tenella infects sheep in many, if not all, countries in the world, including Britain. It is very common, and the cyst is large enough to be seen with the naked eye. Although apparently non-pathogenic, the cyst contains a very powerful toxin ('sarcocystin') which, if extracted and injected to rabbits, is lethal in doses as low as 0·05 mg/kg body weight; given orally, however, the toxin is harmless (which is perhaps fortunate since most non-vegetarians must consume large numbers of cysts of *Sarcocystis* in mutton and beef).

S. lindemanni is the name given to the species found in man. It has been seen only very rarely (on fewer than 20 occasions), but, if normally non-pathogenic, may not in fact be quite so rare as this. Its cyst (Fig. 122) is too small to detect with the naked eye. If the infection is heavy, degeneration of the surrounding muscle fibres and consequent muscular weakness results, with some pain. *S. lindemanni* has usually been found accidentally, while examining muscle sections post-mortem or after a biopsy for some other reason. As a means of diagnosis, a muscle biopsy could be performed, for microscopical examination. Also, a complement fixation test can be used (in man and in other animals), using an antigen prepared from *S. tenella* cysts which have been repeatedly frozen and thawed. Its extreme rarity suggests that *S. lindemanni* is not in fact distinct, but is a species from some other animal (perhaps *S. muris* of rats and mice), which occasionally infects man accidentally.

ADDENDUM

Pneumocystis carinii Delanöe and Delanöe, 1912: a parasite of uncertain taxonomic position

Opinions differ as to whether *P. carinii* is a protozoon or a fungus, though Levine (1961) is adamant that it is the latter. If it is a protozoon, its closest relatives may be the Sporozoa; if a fungus, it is probably related to the yeasts. The parasite causes a disease of man known as atypical interstitial plasma-cell pneumonia, and has been recorded from man, dogs and rodents in North and South America, Europe (including England), Australia and China.

The organisms lie free in the alveoli of the lung. They are spherical, 7 to 10 μ in diameter, and consist of an outer capsule (apparently polysaccharide) and an inner uninucleate protoplasmic body about $1-3 \times 1$ μ in size. The whole organism (capsule and protoplasm) divides by binary fission. Another stage (Fig. 124) is known, in which the capsule is larger (10–12 μ); the inner body is also enlarged and divides into eight uninucleate organisms (about $1 \cdot 5 \times 1$ μ) within the capsule (or 'cyst'). This stage is presumed to be infective, transmission occurring by droplet infection.

P. carinii has been found only rarely in man, being relatively most common in infants (especially premature ones), particularly in central Europe, where epidemics have been recorded in maternity and children's hospitals. *P. carinii* infection has also been seen in a few adults who have been receiving prolonged treatment with corticosteroids, which is known to reduce the capacity to synthesize antibodies. The total number of human infections known is probably not more than a few thousand, almost all in central Europe. The parasite has been found only rarely in animals other than man; it is possible that it may be commoner than has been thought, but rarely pathogenic. Probably either chronically infected adults or domestic animals (or both) may serve as a reservoir from which infants become infected, except during an epidemic, when transmission from child to child undoubtedly occurs.

The pneumonia results from the fact that the parasites, together with the host's plasma cells, block the lung alveoli and bronchioles; also the alveolar walls become thickened and infiltrated with plasma cells. Characteristically no fever is produced. Apart from the clinical picture of an afebrile pneumonia which does not respond to treatment with antibiotics, confirmatory diagnosis is difficult and may depend on finding the organisms in smears prepared from a piece of lung removed at biopsy (or autopsy).

There is at present no specific treatment though the trypanosomicidal compound pentamidine isethionate may be beneficial, and the death rate is high (about 80%). Isolation of suspects is obviously important in preventing the spread of an epidemic through hospitals and similar institutions.

The disease is discussed from the clinicians' viewpoint by White *et al.* (1961); several references are listed in their paper.

10

CNIDOSPORA

For long classified as part of the Sporozoa, this group is now accorded independence as a subphylum. There is some doubt whether its two constituent classes are even as closely related as retention in a common subphylum implies (Lom and Corliss, 1967). They are characterized by the production of resistant, thick-walled spores containing one or more long polar filaments, which are extruded when development in a new host occurs. The spore also contains one or more 'sporoplasms', i.e., germinative cells. All Cnidospora are parasitic with, as far as is known, only a single host, transmission being by ingestion of the spore.

CLASS 1. Myxosporidea
The spore is multicellular and of complicated structure. Of the three orders usually grouped in this class, the Actinomyxida and Helicosporida are little known, rather rare organisms: the former parasitize annelids and sipunculids, while the latter are found in arthropods. They will not be considered further, but more information about them can be obtained from Kudo (1966).

Order Myxosporida
In this, the third and best-known order of Myxosporidea, one or (usually) more polar filaments are present, coiled within special polar capsules (Fig. 125). The spore wall is composed of two or more distinct valves, rather like the shell of bivalved molluscs (e.g. mussels —*Mytilus* spp.), the line where they join being called the sutural line.

All are parasitic in cold-blooded vertebrates, almost exclusively fish, and have a worldwide distribution.

The Society of Protozoologists' Committee (Honigberg et al., 1964) classify the Myxosporida into two suborders, Unipolarina and Bipolarina, depending on whether the polar capsules are grouped at one or both ends of the spore. In this book, however, a classification based on the number of valves forming the shell has been followed, with the position of the polar capsules determining division into superfamilies only (Table 11).

Many Myxosporida produce serious disease in their hosts, particularly those which develop in the viscera. Such diseases may be

TABLE 11
CLASSIFICATION OF THE MYXOSPORIDA
(based on that of Shulman, 1964)

Order Myxosporida
  Suborder 1. Bivalvulina (having 2 shell valves)
    Superfamily 1. Bipolaria (polar capsules at opposite poles of the spore)
                  2. Eurysporea (polar capsules at anterior pole and lying in a plane perpendicular to that of the sutural line)
                    3. Playtsporea (polar capsules at anterior pole and lying in the plane of the sutural line)
  Suborder 2. Multivalvulina (having more than 2 shell valves)

---

fatal, like the 'twist' disease of salmon and trout, which is due to a myxosporidan *(Myxosoma[1] cerebralis)* developing in the cartilage and perichondrium, including that of the skull. Many species infect the skin and muscles of their hosts, and when the latter are food fish the damage done, and the unprepossessing appearance of infected fish (the parasites' cysts are visible to the naked eye), may be of considerable economic importance; 'tapioca' disease of salmon in the Pacific, caused by *Henneguya salminicola* in the muscles, is an example (Kudo, 1966). It also seems probable that heavily infected fish, particularly when organs such as the liver are involved, may grow more slowly (though there is no direct evidence for this). In Lake Victoria, East Africa, virtually all the food fish are infected (Baker, 1963); thus, even quite a small reduction in the growth rate of infected individuals would have a considerable effect on the biomass

---

[1] This generic name is probably a synonym of *Myxobolus;* see Walliker, 1968.

of these fish, and hence on the protein available for food around the lake shore—an area in which protein is a scarce commodity.

Morphology and life cycle

Vertebrates are infected by ingesting spores. These are very variable in shape between the different genera, but all conform to a basic pattern which can be described with reference to one of the genera having less specialized spores, *Myxobolus* (Fig. 125), a member of the superfamily Platysporea (Table 11). Each shell valve is a flattened convex oval, and the two valves are joined in a sutural ridge at the margin of the spore. Within, at the anterior end, are the two polar capsules, each containing a coiled polar filament. Behind the polar capsules is the sporoplasm, a small mass of cytoplasm containing two nuclei which later fuse (see below). The spores of different species vary widely in size, usually between about 10 and 20 μ in length. The spores of some genera (e.g. *Henneguya*) have long posterior processes, possibly as an aid in floating, while others (e.g. *Ceratomyxa*) are elongated laterally, perhaps for the same reason (Shulman, 1964). The life cycle of the group is imperfectly known, though the general pattern seems clear. After a spore is swallowed, it is thought that the polar filaments are everted to serve as an 'anchor' (Fig. 126). The sporoplasm then emerges as a small amoeba which penetrates the gut wall and somehow (possibly via the blood-stream) reaches its appropriate host organ or tissue. Here it becomes a trophozoite and increases in size, with nuclear division, to form a large (often macroscopic) cyst, with a definite limiting membrane. In the forms which inhabit body cavities (coelozoic species), such as gall or urinary bladders or kidney tubules, the cysts lie entirely free. In the forms which inhabit tissues (histozoic species), the cysts are embedded in the tissue (muscle, skin, cartilage, liver, spleen, kidney, etc.) but are apparently extracellular; sometimes they appear to be in or alongside small blood vessels.

Within the cysts, spores develop. Certain cells (sporonts) become differentiated from the syncytial mass. The nucleus of each sporont divides several times to form a sporoblast containing, probably, six nuclei. In most genera, two sporoblasts develop from each sporont: they remain within a common membrane to form what is known as a pansporoblast. The cytoplasm then divides into four uninucleate cells and one which is binucleate. Two of the uninucleate cells form the shell valves, and two form the polar capsules. The binucleate cell becomes the sporoplasm and, at some stage of its development, the two nuclei apparently fuse. This is thought to represent an autogamous sexual process.

When cysts develop within the kidney tubules, urinary bladder and gall bladder or in the skin or muscle of the body wall or gills, it is easy to see how the spores could be liberated by the bursting of the cyst. The spores of species whose cysts develop in tissues such as cartilage or deep muscle, or organs such as the spleen, are presumably dependent upon the death of the host for dissemination. It is possible that the ingestion of an infected fish by a predator, and passage of the spores unchanged through the latter's gut, occurs, though this has not been shown experimentally. Attempts at experimental infection by means of spores have almost always failed, for unknown reasons. Perhaps passage through a predator's gut or some maturation process outside the host is necessary before the spores are infective.

Pathogenesis

The pathological processes involved in myxosporidan infections are little known. Many of those which form visible cysts on the integument and gills of the host probably do it little damage, and the same may well be true of the coelozoic species. Others, however, produce recognizable disease. *Kudoa thyrsites* infects the musculature of the Australian and African barracouta *(Thyrsites atum)*, and produces liquefaction of the surrounding muscle fibres, presumably

---

Fig. 125. *Myxobolus* sp.; spore from liver smear of infected fish (*Tilapia* sp.) From Lake Victoria, Uganda. ×1,500.

126. *Myxobolus* sp.; spore from same source, with polar filaments extended. ×1,500.
(The material from which Figs. 125 and 126 were drawn was originally described as *Myxosoma heterospora*.)

127. Diagram of the structure of a microsporidan spore, greatly enlarged. (From Kudo, Richard R., *Protozoology*, 5th ed., 1966. Courtesy of Charles C. Thomas, Springfield, Illinois, USA.)

128. *Nosema cuniculi*; pseudocyst in macrophage from smear of mouse peritoneal fluid. ×1,500.

129. *N. cuniculi*; liberated spores, some with extruded polar filaments. ×1,500. (Redrawn from an original drawing by Dr R. Lainson.)

Abbreviations: HCN – host cell's nucleus; N – nucleus; P – parasite; PC – polar capsule; PF – polar filament; SPO – sporoplasm; APF – anterior polar filament; BPF – basal portion of the filament; ISM – inner surface membrane; OM – outer membrane; P – polaroplast; PM – polar mass and polaroplast membrane; PPF – posterior polar filament.

Cnidospora

resulting in muscular weakness. *Myxosoma*[1] *cerebralis*, in the cartilage and perichondrium of American salmonid fish, produces distortion of the skeleton and hence of the body of the fish, presumably by some local cytotoxic effect. *Sphaerospora tincae*, a histozoic parasite of the European tench *(Tinca tinca)*, causes distension of the fishes' abdomen which may eventually lead to death by rupture of the abdominal wall.

Myxosporidan infections are diagnosed by identifying the characteristic spores in fresh or stained smears, or in sections, from infected organs. Nothing is known regarding treatment or prevention.

[1] See footnote on p. 138.

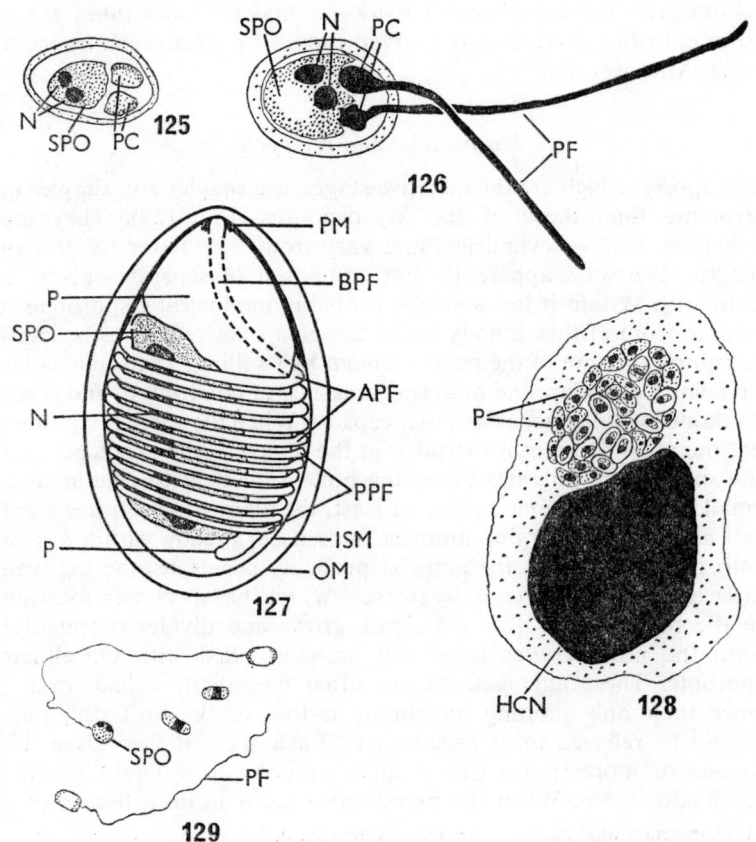

CLASS 2. Microsporidea.

Only one order is recognized within this class—the Microsporida. Its members are differentiated from the Myxosporida by possessing unicellular spores. Microsporida are parasites mainly of arthropods and fish and more rarely of amphibia; however, two species have recently been recorded from mammals, including one record from man (see below). Microsporida are found throughout the world, and some are pathogenic, causing diseases in, for example, economically important insects such as silk-worm (pébrine disease, due to *Nosema bombycis*) and honey-bees (nosema disease, due to *N. apis*). The one known human infection (with *N. cuniculi*) was also pathogenic (see below). Colonies of anopheline mosquitoes, raised in laboratories for experimental work on malaria, sometimes suffer from epizootic disease of the larvae caused by *Thelohania legeri* or other Microsporida.

Morphology and life cycle

The spores, which are the infective stages, are smaller and simpler in structure than those of the Myxosporida (Fig. 127). They are spherical, oval or cylindrical and vary from 5 or fewer to 10 μ in length. The wall, apparently not composed of separate valves, is chitinous. Within it lies a single, probably uninucleate, sporoplasm and, in front of this, a body called the polaroplast which is believed to cause extrusion of the polar filament by swelling. The single polar filament is coiled around both sporoplasm and polaroplast, and is not contained within a separate polar capsule. When the spore is swallowed, the polar filament is extruded in the lumen of the host's gut and the sporoplasm migrates along the hollow filament to emerge as a small amoeba. In some species at least, the filament penetrates a gut cell and so the emerging amoeba finds itself already within a host cell: in other species, the amoeba probably penetrates the gut wall after emergence. It eventually (somehow) reaches its chosen location in the body, where it enters a cell, grows and divides repeatedly[1] until the now very enlarged cell becomes filled with unicellular sporonts. These aggregations are often incorrectly called 'cysts'; since their only limiting membrane is that of the host cell, they should be referred to as pseudocysts. Each sporont then gives rise to one or more spores (the number differing in different genera; see Kudo, 1966). When the pseudocysts occur in deep tissues of a

[1] It is uncertain whether this process should be regarded as schizogony or not.

vertebrate host, the release of the spores is presumably dependent upon the host's death; in other instances, however, they may be liberated directly from a rupturing skin pseudocyst, or into various body cavities which are connected with the outside world (e.g. kidney tubules of vertebrates). Release from infected insects is probably usually dependent on the death of the host, though some species are known to be transmitted via the egg to the larvae (e.g. *N. bombycis* of the silk-worm).

Although all Microsporida have unicellular spores, as described, in one genus with a single species *(Telomyxa glugeiformis)* the spores are stuck together in pairs by the posterior ends so that, at first sight, each spore seems to have two sporoplasms and polar filaments. Because of this, *Telomyxa* is usually classified in a separate suborder, the Dicnidina; all other Microsporida are grouped in the suborder Monocnidina, in various familes.

(The above account is a 'traditional' one; for a stimulating, if more unconventional account of microsporidan morphology and life history, see the paper by Sprague and Vernick, 1968.)

Pathogenesis

Little is known about the pathogenesis of microsporidan infections, though many (especially among insects) are fatal. Infected cells usually increase enormously in size, as do their nuclei, and are presumably rendered unable to perform their natural function. Thus, if an organ is heavily infected it may become virtually useless to the host and this, if the organ is a sufficiently important one, may lead to the death of the host. For example, *Plistophora myotrophica* of the toad *(Bufo bufo)* infects the skeletal muscles, which atrophy. If heavily infected, the toad becomes so weak that it dies from exhaustion, being presumably unable to feed (Canning *et al.,* 1964). Sometimes these parasites exert a harmful effect on organs other than those which they infect. For instance *N. apis* infects only the gut cells of the honey-bee, but the ovaries of infected queens degenerate (Kudo, 1966).

Microsporidan infections are diagnosed by recognizing the spores in sections or smears of infected organs. Little is known about their treatment or prevention, though the antibiotic Fumagillin is used to control *N. apis* of bees.

Microsporida in mammals

Only two species of Microsporida are known to infect mammals. One of these is *Thelohania apodemi*, which has been found in the brains of voles *(Apodemus* sp.*)* in France (Doby *et al.*, 1963), and the other is *Nosema cuniculi*.

N. cuniculi has been known since 1923 under the name *'Encephalitozoon' cuniculi*, as a parasite of wild and domestic rodents and lagomorphs, including laboratory colonies. The discoverers of this organism, Levaditi, Nicolou and Schoen, suggested that it might be a microsporidan, but most people thought that it was probably related to *Toxoplasma*. However, a careful morphological study by Lainson *et al.* (1964) showed that *'Encephalitozoon'* was unquestionably a microsporidan of the genus *Nosema* with small, oval spores measuring about $2 \cdot 5 \times 1 \cdot 5$ μ (Figs. 128, 129). In laboratory animals the parasite usually produces chronic, symptomless infections in which it can be found only in pseudocysts in the brain. It may, however, cause acute generalized infections, in which pseudocysts can be found within monocytes and other lymphoid-macrophage cells in all viscera and in the peritoneal fluid; from such infections mice, at least, usually die. There has been one definite record of *N. cuniculi* from man (see Lainson *et al.*, 1964 for reference): parasites were found in the cerebro-spinal fluid and urine of a Japanese boy severely ill with an encephalitis. It may be that *N. cuniculi* infects man more commonly, but usually produces a chronic symptomless infection and it may be that some acute infections in the past have been mistakenly ascribed to other causes.

Little is known concerning the transmission of *N. cuniculi*, and nothing about the treatment or prevention of infection. In the laboratory, the parasite has been transmitted from rabbit to rabbit by ingestion of urine, and the finding of the parasites in the urine of the Japanese boy is interesting in this light. Perhaps such human infections as do occur arise mainly from eating food contaminated with the urine of infected rats and mice.

11

PARASITIC CILIATES

The ciliates constitute a class apart amongst the Protozoa. They are a vast group and a vast subject, and will be dealt with here only very briefly. A detailed taxonomic and morphological account of the ciliates, parasitic and free-living, can be obtained from the 'bible' of ciliatology, Corliss's monograph (1961); a good general account is given by Mackinnon and Hawes (1961); and a useful key is given by Corliss (1959): but the classification used in these works has been slightly revised by the Committee of the Society of Protozoologists (Honigberg *et al.*, 1964), whose version is followed here. The great majority of ciliates are free-living, parasitism (in its widest sense) having apparently arisen independently amongst the various groups. Indeed, within the single genus *Tetrahymena* (subclass Holotrichia, order Hymenostomatida), all gradations can be seen ranging from species which are entirely non-parasitic through those which can live equally well both free or within the body cavities of insects etc. (facultative parasites), to species which are almost entirely parasitic with only occasional periods of 'free' existence during their life cycle (Corliss, 1960).

A few orders consisting entirely of parasitic forms have evolved amongst the subclass Holotrichia: these are the Apostomatida, containing species which are ectoparasitic on crustacea, the Astomatida (endoparasitic in oligochaete annelids), and the Thigmotrichida (endo- and ecto-parasitic in or on bivalved molluscs). Some members of the orders Gymnostomatida, Trichostomatida, Chonotrichida and Hymenostomatida are endo- or ecto-parasitic (the Hymenostomatida includes *Ichthyophthirius,* a common and sometimes harmful ectoparasite of fish). Also, in the subclass Spirotrichia, the order Entodiniomorphida is a very specialized group inhabiting the stomach of herbivores (see below, pp. 149–151); some of the Heterotrichida are also parasitic. Many of the sessile forms comprising the subclasses Peritrichia and Suctoria are attached to various aquatic animals, and may sometimes be injurious to their hosts, e.g. *Trichodina* (subclass Peritrichia, order Peritrichida, suborder Mobilina) on fish.

The only ciliate of medical or veterinary importance is *Balantidium coli* (subclass Holotrichia, order Trichostomatida), which is described on pp. 145–149 below.

All ciliates are characterized by the possession of cilia (p. 29) during part or all of their life cycle. The basal bodies of these cilia, together with an associated complex of fibrils, are present throughout the life cycles of all species (see Mackinnon and Hawes, 1961). Ciliates almost always have nuclei of two types—a macronucleus and one or more micronuclei within each individual—the former being concerned with the day-to-day somatic functioning of the organism, the latter solely with sexuality. Sexual reproduction takes the form of the unique process known as conjugation. Basically, this process (described in detail by Mackinnon and Hawes, 1961, pp. 293–300) occurs as follows. Two ciliates pair and become united by the temporary dissolution of the contiguous parts of their pellicles; their macronuclei disintegrate and their micronuclei divide twice, one of these divisions being meiotic; three of the resulting micronuclei degenerate, while the fourth divides once more to produce two gametic nuclei. One of the gametic nuclei (the 'male') from each member of the conjugating pair migrates into the other member of the pair, where it fuses with the stationary ('female') nucleus. The zygotic nucleus then divides twice and the pair of ciliates separates; both individuals then divide, each daughter receiving two nuclei, one of which becomes polyploid and forms the new macronucleus while the other constitutes the new micronucleus.

Genus *Balantidium* Claparède and Lachmann, 1858
This genus is classified in the subclass Holotrichia, order Trichostomatida. It contains the only species of ciliate which infects man, *B. coli*. About 50 species of *Balantidium* have been described, mainly from primates and amphibia, but some are probably synonyms; other animals which have been recorded as hosts of *Balantidium* include pigs, sheep, guinea-pigs, camels, opossums, ostriches, fish, cockroaches and other insects, flatworms, coelenterates, as well as (rather doubtfully!) plants.

Parasitic ciliates

Specimens of *Balantidium* for study may be obtained from pig faeces or the rectal contents of frogs (killed with ether or chloroform). Another ciliate which may be found in the rectum of frogs is *Nyctotherus cordiformis* (subclass Spirotrichia, order Heterotrichida, suborder Heterotrichina). The genus *Opalina* (p. 75) also occurs there; this organism is much larger than *Balantidium*. Another species of *Nyctotherus*, *N. ovalis*, may be found in the intestine of cockroaches. *Nyctotherus* can be distinguished from *Balantidium* by the lateral position of its 'mouth' (peristome).

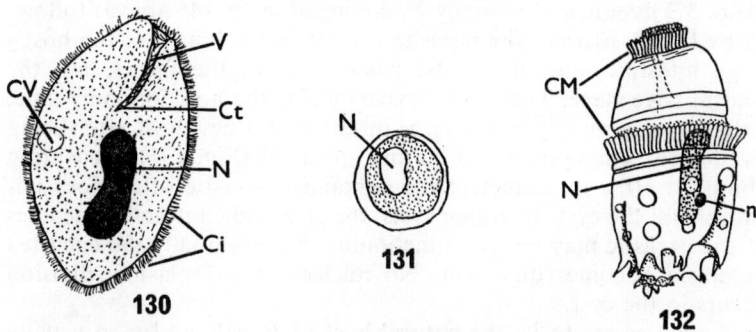

Fig. 130. *Balantidium coli*; trophozoite of an amicronucleate strain from culture. ×500.

131. *Bal. coli*; cyst in fresh preparation of pig faeces (unstained). ×250.

132. *Ophryoscolex caudatus* from rumen of ox. ×200. (Redrawn and slightly modified from Levine, 1961, fig. 38E.)

Abbreviations: Ci – cilia; CM – ciliary membranelles (fused cilia); Ct – cytostome; CV – contractile vacuole; N – macronucleus; n – micronucleus; V – vestibule.

Balantidium coli inhabits the large intestine (caecum and colon) of man, apes, monkeys and pigs; it has also been recorded occasionally from dogs, rats, sheep and cattle. The pig is the commonest and probably the main natural host. *B. coli* has been recorded from all parts of the world. In man (and other primates) *B. coli* causes a disease called balantidiosis or balantidial dysentery. In pigs it is apparently usually non-pathogenic. The ciliate is relatively large and ovoid (Fig. 130), measuring about 60–70 × 40–60 μ. (Levine, 1961, gives its measurements as 30–150 × 25–120 μ, but such a range is exceptional.) The entire surface is covered with cilia arranged in longitudinal, slightly spiral rows. At the front is a deep groove or vestibule, lined with slightly longer cilia, which leads to the mouth or cytostome. The animal feeds by drawing a current of water down the vestibule, by means of the cilia lining the latter. Any food particles carried in this current are ingested through the cytostome in small food vacuoles, which then circulate along a defined path through the ciliate's body while their contents are being digested. Eventually any undigested remnants are expelled through a permanent pore or cytopyge near the organism's hind end. The parasite has two contractile vacuoles and a macronucleus and micronucleus.

B. coli reproduces asexually by transverse (i.e. homothetogenic; see p. 37) fission, and sexually by conjugation (p. 146 above) followed by binary fission. The parasite encysts in the lumen of the host's large intestine, and the cysts, passed out in the faeces, are the transmissive stage. They can survive outside the host for some time, hatching presumably in the large intestine of a new host after being swallowed. The cysts are large and spherical (Fig. 131), measuring about 50–60 μ in diameter; the contained parasite often does not entirely fill the cyst. In young cysts, the cilia and contractile vacuoles of the parasite may be seen functioning, but eventually the encysted organism becomes quiescent. No nuclear or cytoplasmic division occurs in the cyst.

The pig seems to be the natural host of *B. coli,* and man usually becomes infected from a pig, by swallowing cysts passed out in the latter's faeces. Consequently, most human infections are in country-dwellers and farmworkers. Human infections are, however, rare, the infection rate being less than 1%; in pigs by contrast, it is between 21 and 100% (Levine, 1961).

In man, the ciliates not only inhabit the lumen of the large intestine but they also invade the intestinal wall, penetrating the mucosa and submucosa and producing lesions like those caused by invasive *E. histolytica* (p. 82). The irritation produced by the ciliates leads to diarrhoea, and as the lesions progress blood vessels become eroded and bloody dysentery results. In sections of infected areas of intestine, the parasites are easily recognized because of their large size and large macronuclei; they are often found in groups known as 'nests' or (in Latin) 'nidi' (singular, 'nidus'). In the pig the ciliate is normally non-invasive and, therefore, non-pathogenic. Apparently it cannot penetrate the intact intestinal mucosa, but if the latter is damaged by infection with bacteria such as *Salmonella,* or some other cause, invasion can occur resulting in lesions similar to those seen in man. Whether the ciliate also needs help to penetrate the human mucosa is not known—quite possibly it does.

Balantidium infection can be diagnosed easily by finding the ciliates, free or encysted, in faecal specimens (pp. 152–155). The parasites will grow in the ordinary culture media for intestinal amoebae (pp. 155–156), but recourse to this aid is seldom necessary for diagnosis.

In man, the infection can be treated with various drugs, including carbasone (p. 72), di-iodohydroxyquin and the arsenical stovarsol; the antibiotics chlortetracycline and oxytetracycline are also said to be effective. Treatment of pigs is unnecessary as the parasite is usually non-pathogenic. Prevention of infection in man is simply a matter of elementary sanitary precautions and the avoidance of a diet which includes pig faeces.

THE RUMEN CILIATES

What Levine (1961) rightly calls a 'bewildering variety' of ciliates inhabits the rumen and reticulum of ruminants; another, and equally bewildering group, lives in the caecum and colon of equids (Table 12). Many of these forms have a most complicated structure, especially those in the family Ophryoscolecidae (subclass Spirotrichia, order Entodiniomorphida), many of which have pellicular skeletal plates (of polysaccharide material) and a very reduced ciliature (Fig. 132). Many of these and other genera of rumen ciliates are illustrated by Levine (1961), who also gives descriptions of all the genera involved. As far as is known, none of these forms produces cysts. The rumen-dwellers are killed when they enter the host's abomasum, and transmission occurs when young animals feed on hay and grass contaminated with the saliva of older, infected animals; the saliva contains ciliates regurgitated during cud-chewing. Less is known about the forms living in the large intestine of equids, but these too are presumably transmitted orally on food contaminated with faeces containing the ciliates. These ciliates are known from all parts of the world, from cattle, sheep, goats, camels, reindeer, water buffalo, elk and antelope.

Relationship to the host

Little is known about the relationship of the ciliates of equines to their hosts—probably they are all harmless commensals. The host-relationships of the rumen ciliates have been much more studied; apparently none is harmful, most are harmless commensals, and a few are of definite value to their hosts. Unlike the symbiotic flagellates of termites (p. 74), however, none of these forms is essential to its

TABLE 12

CILIATES LIVING IN THE RUMEN AND RETICULUM OF RUMINANT HERBIVORES, AND IN THE CAECUM AND COLON OF EQUIDS

Hosts	Subclass	Order	Suborder	Family	Genera
RUMINANTS	Holotrichia	Gymnostomatida	Rhabdophorina	Buetschliidae	*Buetschlia*
		Trichostomatida	—	Isotrichidae	*Isotricha, Dasytricha*
	Spirotrichia	Entodiniomorphida	—	Ophryoscolecidae	*Ophryoscolex, Entodinium, Diplodinium* and 15 other genera
EQUIDS	Holotrichia	Gymnostomatida	Rhabdophorina	Buetschliidae	13 genera (not *Buetschlia*)
		Trichostomatida	—	Blepharocorythidae	*Blepharocorys, Charonina, Ochoterenaia*
				Paraisotrichidae	*Paraisotricha*
	Suctoria	Suctorida	—	Acinetidae	*Allantosoma*
	Spirotrichia	Entodiniomorphida	—	Cycloposthiidae	7 genera

host. This subject has been reviewed by Hungate (1955) and Oxford (1955) and summarized by Levine (1961, pp. 359–62). It will be dealt with only briefly here.

Diplodinium and some other Spirotrichia can digest cellulose; there is some doubt as to whether they do this by virtue of their own enzyme (cellulase) or that of symbiotic bacteria. These and many other genera also digest starch and soluble carbohydrates. Thus they must help the host in breaking down these substances, especially the cellulose (ruminants, like all mammals, do not produce a cellulase). However, the ruminants can live quite satisfactorily without the ciliates, cellulose digestion then being performed by the rumen bacteria alone. Other ways in which the ciliates must benefit the host are in storing carbohydrates and digesting them gradually, thus making available to the host a more regular supply of the volatile fatty acids into which the carbohydrates are broken down (all rumen ciliates are obligate anaerobes).

Although the ciliates convert some of the host's food into their own protein, this is not lost to the ruminant but becomes available to it when the ciliates die. This in fact benefits the host since the ciliate protein is of higher quality nutritionally than that provided by the plants on which it feeds or by the rumen bacteria. Hungate (1955) has estimated that at least 100 g of protein are provided daily by the rumen ciliates in an average ox—about one-fifth of the ruminant's total daily protein requirement. Estimates of the numbers of ciliates present in the rumen vary widely, but they are all immense: Hungate (1955) suggests about 100,000 per ml, which, in an ox with 100 kg of rumen contents, would amount to approximately 10,000 million ciliates (three times the human population of the world in 1966).

12

TECHNIQUES FOR INTESTINAL PARASITES

Parasitic Protozoa which inhabit the alimentary canal of their hosts may be obtained for study either post-mortem or by the collection of faecal specimens. Usually the latter contain only encysted forms, unless the stool is diarrhoeic, when trophozoites of amoebae, flagellates and ciliates may be found. Many free-living Protozoa (coprozoic forms) multiply rapidly in old, moist faeces, and care has to be taken not to confuse them with parasitic forms in samples which are not fresh. Specimens which cannot be examined immediately should be refrigerated at 4°C if possible, and kept in closed vessels.

Specimens may also be preserved by emulsification in 10% formol-saline (4% HCHO in 0.9% saline).[1] Faecal samples may be examined either directly or after concentration.

EXAMINATION OF FAECAL SPECIMENS

Direct

For direct examination, a small portion of the specimen (about the size of a large pin's head) is taken on a matchstick or swab stick and thoroughly dispersed in a drop of fluid on a microscope slide, covered with a coverslip and examined. If it is thought that trophozoites may be present, one preparation should be made in 0.9% saline; two other similar preparations should always be made, whether the presence of trophozoites is suspected or not—one in 1%

---

[1] 'Formol' or 'formalin' is a 40% aqueous solution of the gas formaldehyde (HCHO).

Techniques for intestinal parasites

aqueous eosin solution and one in double-strength Lugal's iodine solution (4% potassium iodide plus 2% iodine in distilled water). Practice is needed in making these preparations—if they are too thick it will be impossible to examine them microscopically, and if they are too thin, parasites may not be seen unless present in very large numbers. The saline preparation (if made) should be examined microscopically at a total magnification of about ×100, using phase-contrast illumination if possible or with the intensity of illumination reduced by partially closing the substage iris diaphragm; any motile organisms seen can be examined further at a magnification of about ×400. The eosin preparation should be examined similarly. If it is correctly made (without an excess of eosin), living protozoan cysts (and helminth eggs) will appear as small unstained (white) objects against the pink background of eosin solution and the red-stained debris. They can thus be readily detected and subjected to further examination at magnifications of ×400 and (using an oil immersion objective) ×500–×1000, if necessary. Little detail can usually be seen in these unstained specimens, except for the chromatoid bodies which may be seen in cysts of *Entamoeba* spp. These should be searched for carefully and, if present, their shape noted (see pp. 78–80 above). The shape and size of any cysts seen should also be noted.

If cysts have been seen during this preliminary examination of the eosin preparation, that suspended in iodine should be examined. Here the cysts will be less obvious, as almost everything is stained yellow-brown by the iodine, and it is better to commence the examination at a higher magnification (×400); the oil-immersion objective can then be used to study individual cysts in more detail. Iodine stains the nuclei of cysts, and their number and structure should be noted; also, vacuoles containing glycogen, if present, will be conspicuous as their contents stain a deep golden-brown colour. (Chromatoid bodies are *not* clearly seen in cysts stained with iodine.)

After concentration

Scanty infections are easily missed if only direct examinations of faecal specimens are made. If possible a method of concentrating the cysts (and helminth eggs) in a specimen should be used (these methods, however, kill trophozoites).

(a) A simple method of concentration is the suspension of a small portion of the faeces (about 15 ml = 1 in^3) in a 33·1% aqueous solution of zinc sulphate ($ZnSO_4 \cdot 7H_2O$), using a small pestle and mortar;

place the suspension in a glass cylinder measuring about 5 × 2 cm (2 × ¾ in.)—which must be filled to the brim—and then place a coverslip over the mouth of the cylinder, touching the fluid; after 20–30 minutes, remove the coverslip and place it (fluid side down) on a microscope slide. Any cysts present should have floated up in the zinc sulphate solution and been removed in the film of liquid attached to the coverslip. If cysts are present, they can subsequently be stained by placing a drop of double-strength Lugal's iodine solution (p. 153) alongside the coverslip and drawing the solution beneath the latter by applying a piece of filter paper to the opposite edge of the coverslip.

(b) A better method of concentration is the formol-ether technique, which requires the use of a centrifuge. 1–2 g of faeces is emulsified in 10 ml of 10% formol-saline (p. 152) and strained through a wire sieve (16 mesh/cm = 40 mesh/in.) into a small centrifuge tube; formol-saline is added to bring the fluid level to about 2·5 cm (1 in.) of the top of the tube; about 3 ml of ether is then added, the tube is shaken vigorously and then centrifuged, the speed being increased gradually to a maximum of 2,000 rpm (700 × g) after 2 minutes and the centrifuge then switched off. When the tube has come to rest, the fatty debris at the interface of the formol-saline and the ether is loosened from the wall of the tube with a matchstick or swabstick, and the debris plus supernatant fluid is poured off and discarded. The deposit is then re-suspended in the small drop of fluid remaining in the tube, and removed in a Pasteur pipette; half is added to a drop of Sargeaunt's stain (0·2 g malachite green dissolved in 3 ml of 95% ethanol; 3 ml acetic acid is added and the volume made up to 100 ml with distilled water) on a microscope slide. This stain colours chromatoid bodies and nuclei dark green. The other half of the resuspended deposit is added to a drop of double-strength Lugal's iodine solution to reveal the presence of any glycogen vacuoles.

The oocysts of intestinal coccidia are also concentrated by these two methods. However, in order to identify species it is often necessary to keep the oocysts *in vitro* at room temperature for 2–3 days for sporulation to be completed, which cannot be done after formol-ether concentration as this kills the parasites. Zinc sulphate solution is also likely to be harmful. The simplest method to concentrate coccidian oocysts is to suspend the faeces in a saturated aqueous solution of sodium chloride and centrifuge the suspension at 1,500 rpm (about 400 × g) for 2 minutes. The surface layer of solution, containing the oocysts, is at once pipetted into 3 or 4 times its volume of water, which dilutes the saline sufficiently to allow the oocysts to

Techniques for intestinal parasites

sink (by gravity or after further centrifugation); the sedimented oocysts are finally resuspended in 2% aqueous potassium dichromate solution and kept in a shallow layer of this solution in a Petri dish at room temperature until sporulation has occurred (as judged by the microscopical examination of a drop of the suspension).

PERMANENT PREPARATIONS

Permanently stained preparations of intestinal Protozoa may be made by spreading (with a matchstick or swabstick) a thin film of the faecal matter on to a coverslip and fixing it before drying in Schaudinn's solution (60 ml of a saturated solution of mercuric chloride in 0·9% saline, 30 ml of ethanol and 10 ml of acetic acid) for 20–30 minutes. The smear is then briefly washed in 70% ethanol containing a little (3–5%) triple-strength Gram's iodine solution (to remove the mercuric ions introduced with the fixative) followed by five minutes immersion in 70% ethanol containing 3–5% of 5% aqueous sodium hyposulphite. It is then briefly passsed through fresh 70% ethanol, 90% ethanol and pure ethanol (5 minutes, to harden the parasites), and rehydrated through 70%, 50% and 30% ethanol to distilled water. The preparation is then placed in tungstophosphoric acid haematoxylin stain[1] for 12 hours or longer: overstaining is impossible. The preparation is briefly washed in water, dehydrated through 30%, 50%, 70%, 90% and pure ethanol, cleared in xylene and mounted (film downwards) on a slide with Canada balsam.

CULTIVATION in vitro

Some intestinal Protozoa grow readily in vitro at 37°C in simple media such as Dobell and Laidlaw's 'HSre + S', prepared as follows.

Sufficient sterile horse serum is poured into sterile cotton-plugged test-tubes so that, when the tubes are held at an angle of about 40°, the serum runs 3 or 4 cm up the tube. While lying at this angle, the tubes are heated at 80°C in a water bath for 60–70 minutes to coagulate the serum.

Fresh hen's eggs are cleaned and sterilized by washing in alcohol and a small hole is cut with sterile scissors in the blunt end of the egg; each egg is then inverted over a flask containing sterile Ringer's[2]

---

[1] 0·1 g haematoxylin is dissolved by heat in a little distilled water, cooled, and the volume made up to 80 ml with distilled water; 20 ml of 10% aqueous tungstophosphoric acid (analytical reagent grade) is then added. This solution must be allowed to 'ripen' for several months before use (the 'ripening' can be accelerated by adding 10 ml of 0·25% aqueous potassium permanganate).

[2] Formula in g/l: NaCl, 9·0; KCl, 0·42; $CaCl_2$, 0·24; $NaHCO_3$, 0·2; glucose, 1·0.

solution and, by puncturing the narrow end of the egg, the albumen is allowed to run out of the hole into the flask (4 eggs are used for each litre of saline). If not prepared aseptically, this solution must be sterilized by filtration.

The albumen-Ringer solution is added to the test-tubes containing the slope of coagulated serum, just to cover the latter. The completed tubes should be incubated at 37°C for 24–48 hours to check their sterility, and may be stored in a refrigerator for up to one week before use. Just before use, a small amount of rice starch (sterilized by heating to 160°C for 1 hour) is added to each tube.

For inoculation, a little faecal matter is introduced, by means of a bacteriological wire loop, to the bottom of the overlay. The tubes should be warmed to 37°C before inoculation, incubated at 37°C and examined and subinoculated daily for the first few days; established cultures need be sub-inoculated only every 2–4 days. Subinoculations are usually made with a Pasteur pipette: the base of the slope is scraped to dislodge any amoebae, and then 0·5–1 ml of the sediment at the bottom of the tube is transferred to a new tube.

If this medium is being used for diagnosis, it is advisable to subinoculate for 2 or 3 days even if no amoebae are seen, since it may take a few days for reasonable numbers of trophozoites to develop. Most intestinal flagellates of mammals (except *Giardia*), *Balantidium coli* and amoebae of the *Entamoeba histolytica* group, grow readily in this medium. Further techniques of cultivation are given by Taylor and Baker (1968).

HISTOLOGICAL SECTIONS

Tissues (e.g. intestinal wall) infected with Protozoa such as *E. histolytica*, *Bal. coli* or *Histomonas meleagridis* can be fixed, embedded and sectioned by the usual histological procedures, and stained with haematoxylin and eosin (details of the method can be obtained from standard histology text-books, e.g. Drury and Wallington, 1967): alternatively, the Giemsa method for sections can be used (see p. 162 below).

AVAILABILITY OF MATERIAL FOR STUDY

Obtaining parasites of man for study may be difficult, but acceptable substitutes can often be found in domestic and, if available, laboratory animals.

Rhesus monkeys may harbour most of the common human intestinal parasites, but a more readily available substitute host is the pig. Pig faeces will usually contain cysts of *E. suis* (which resemble those of *E. histolytica* except that they are uninucleate), *Bal. coli* and *Iodamoeba buetschlii*. Species of *Giardia* and *Trichomonas* can be obtained from laboratory rodents, especially hamsters and guinea pigs; *Hexamita* may also be present. Intestinal coccidia of the genus *Eimeria* are easily obtainable from laboratory or domestic rabbits and chickens; young ones are more likely to be heavily infected. *Isospora* spp. are commonly found in wild passerine birds. The rectum of frogs usually contains a ciliate called *Nyctotherus,* and sometimes a species of *Balantidium* (not *Bal. coli*) and *Opalina*. Examination of one's own faeces may sometimes be rewarding—and surprising.

13

TECHNIQUES FOR TISSUE PARASITES

Protozoa living in the blood and other tissues of their hosts can be obtained for study either after the death of the host or, sometimes, depending on the tissue which they inhabit, during life by removing a small portion of the appropriate tissue. The latter procedure should be attempted only by qualified workers with access to suitable equipment and anaesthetic and operating facilities, and care must be taken that it does not infringe the legislation of the country in which it is being done.

BLOOD FILMS

Small quantities of blood can be simply obtained (subject to the above provisions) from small rodents by pricking the tail near its tip with a needle, or by cutting off its extreme tip with scissors. For larger mammals, pricking an ear vein is often satisfactory; with man, the finger tip may be pricked with a sterile needle (having sterilized the skin with alcohol and allowed it to dry). Having obtained a drop of blood, films may be prepared in two ways.

Thin blood films

The blood drop is placed near one end of a microscope slide; another slide is then placed with its narrow edge touching the drop and inclined so that the drop runs along the acute angle (about 30°) formed by the two slides; the inclined slide is then pushed along the horizontal one in the direction of the *obtuse* angle between the two slides, so that the blood drop is pulled (*not* pushed) along the horizontal slide (Fig. 133). The resulting film is then dried rapidly by waving it in the air (*not* by heating), and should be fixed and stained

133(a)

133(b)

134(a)

134(b)

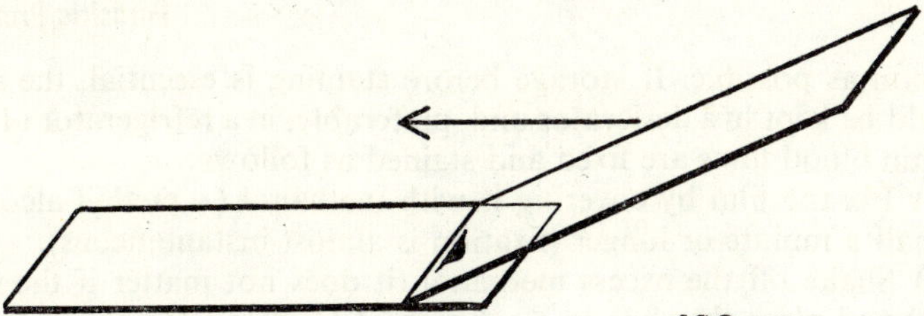

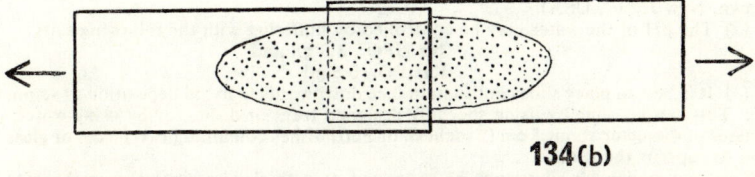

Fig. 133. Diagram illustrating the preparation of a thin blood film.
134. Diagram illustrating the preparation of a brain smear. (See text for details.)

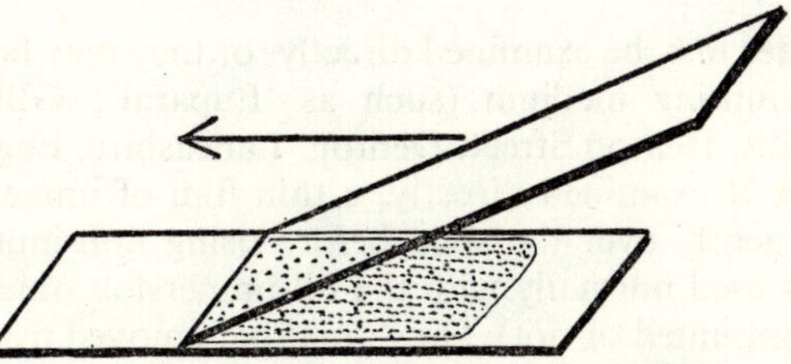

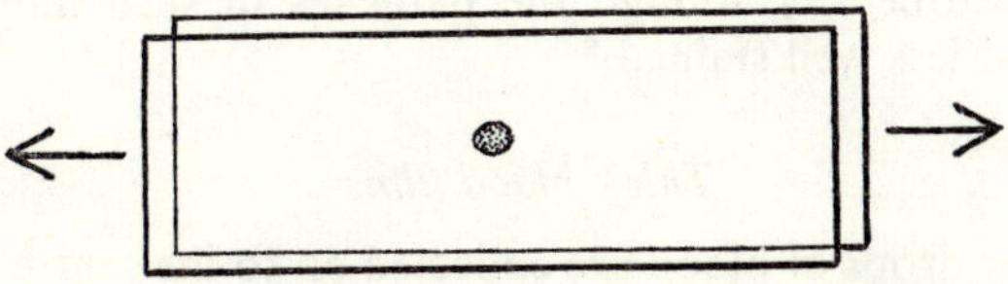

as soon as possible. If storage before staining is essential, the slide should be kept in a desiccator and, preferably, in a refrigerator (4°C).

Thin blood films are fixed and stained as follows:

(1) Fix the film by covering it with methanol (=methyl alcohol) for half a minute or longer (fixation is almost instantaneous).

(2) Shake off the excess methanol (it does not matter if the film dries) and place the slide in a solution of Giemsa's stain (1 volume) in distilled water at pH 7·2 (10 volumes) for 50–60 minutes.[1]

(3) Remove the slide from the stain and wash it by flooding *very briefly* (not longer than 1 second) under a tap.

(4) Place the slide in an upright position to drain and dry (do *not* use heat).

Such slides may be examined directly, or they may be mounted in a neutral mounting medium (such as 'Euparal'; GBI (Labs) Ltd, Heaton Mills, Heaton Street, Denton, Lancashire, England) beneath a coverslip. If examined directly, a thin film of immersion oil must be spread gently over the slide before using non-immersion objectives (oil is used normally with the oil-immersion objective, whether the film is mounted or not). Oil should be removed from unmounted films after examination by washing them in xylene (neutralized if necessary with calcium carbonate).

Thin films result in the best morphological preservation of parasites. However, if the organisms are very scanty, searching for them in a thin film may be very laborious. If mammalian blood is being studied, this difficulty can be partially overcome by making a thick film (see below), though the parasites in such films may be distorted and less well stained.[2]

Thick blood films

Three or four drops of blood are collected on to the centre of a slide, and spread with a needle (or the corner of another slide) into a

[1] *Notes:* (1) Not all brands of Giemsa's stain are equally satisfactory: three reliable brands are 'Revector' (Hopkin and Williams, Ltd, Chadwell Heath, Essex, England) and those of the National Aniline Division of Allied Chemical and Dye Corporation (40 Rector Street, New York 6, NY, USA) and the Fisher Scientific Company (Fair Lawn, New Jersey, USA).

(2) The pH of the water may be controlled by buffering with the following salts:
Na_2HPO_4, 3·0 g/l and
KH_2PO_4, 0·6 g/l

(3) It is best to place slides in the stain face downwards to avoid deposition of scum, etc. This can be done by using specially designed trays or dishes, or by using watch-glasses of diameter about 5 cm (2 inches), or Petri dishes containing two pieces of glass rod to support the slides.

[2] Non-mammalian blood cannot be examined as a thick film since the erythrocyte nuclei would interfere.

circular area about 1–1·5 cm (½ in) in diameter. The film should be sufficiently thick for it to be just possible to read small print (e.g. the figures on a wrist-watch) through it. In order to make this thick preparation sufficiently transparent for microscopical examination, the blood (after thorough drying) is lysed in a hypotonic solution. This releases the contained haemoglobin from the erythrocytes, leaving any parasites more or less undamaged. It is, therefore, essential that thick films should *not* be fixed, since fixation prevents lysis. The staining procedure is as follows (the films must be thoroughly dry—about 12 hours drying at room temperature, or 2–4 hours in a desiccator or at 37°C—*not* more—is advisable).

(1) Immerse the film in 0·5% aqueous methylene blue for 1 second (this is not essential, but gives better results).

(2) Rinse it briefly (and gently[1]) in tap water.

(3) Place the slide face downwards in Giemsa's stain (prepared as described above for thin films) for 30 minutes (not longer).

(4) Continue as described for thin films (paragraph 3 et seq., p. 160 above): thick films are examined similarly to thin films.

For parasitic Protozoa, blood films are usually examined only with a ×100 oil-immersion objective, though preliminary scanning with a ×40 'high-dry' (i.e. non-immersion) or (if available) ×50 oil-immersion lens may be used to locate larger forms (e.g. *Leucocytozoon,* some trypanosomes).

Further information on the making and staining of blood films is given by Shute and Maryon (1966).

TISSUE IMPRESSION SMEARS

Small pieces of tissues such as lung, heart, liver, kidney and spleen may be removed from dead animals. The cut surface is dabbed several times on filter-paper or blotting-paper to remove most of the blood, and then pressed (not smeared) once on to a slide (several dabs may be made on one slide). After drying, such smears are fixed, stained and examined in the way described above (p. 160) for thin blood films.

Smears of brain may be made by placing a small piece (pin's head size) near one end of a slide, crushing it by placing another slide on it, and spreading it out by sliding the two slides longitudinally over one another (Fig. 134). Smears of bone marrow are usually made by spreading a small fragment on a slide with the aid of a needle.

[1] Thick films must be rinsed *very gently*, either by dipping into water or holding under a *slowly* running tap. They come off the slide easily if handled too roughly.

HISTOLOGICAL SECTIONS

Viscera suspected of harbouring parasites can be fixed, embedded and sectioned by standard histological procedures. The sections may then be stained with haematoxylin and eosin (see, e.g., Drury and Wallington, 1967) or with Giemsa's stain. Small pieces of tissue for staining with Giemsa should be fixed for about 3 hours in Carnoy's fluid (ethanol, 6 volumes; chloroform, 3 volumes; acetic acid, 1 volume) and washed in 90% ethanol before dehydration, clearing and embedding in the usual way. The staining procedure is as follows (Bray and Garnham, 1962).

(1) After hydration of the sections, place them in a mixture of 1 volume of Giemsa's stain (see note (1) on p. 160), 1 volume of methanol and 10 volumes of distilled water at pH 7·2 (see note (2) on p. 160) for 1 hour.

(2) Remove the slides from the stain and wash them in tap water.

(3) Differentiate the sections by pouring a 15% solution of colophonium resin in acetone on to the wet slides, rocking the latter to and fro for 3–10 seconds, and pouring off the solution. Repeat this process 2 or 3 times until the blue-green dye no longer streams out in the differentiating solution.

(4) Wash the slides rapidly in a mixture of xylene (7 volumes) and acetone (3 volumes), followed by several washings in pure xylene.

(5) Finally mount the slides in a *neutral* medium (e.g. 'Euparal'; see p. 160), beneath a coverslip.

CULTIVATION *in vitro*

Many species of *Trypanosoma* other than the Salivaria, *Leishmania* (and other Trypanosomatidae) can be grown in test-tubes plugged with non-absorbent cotton wool or screw-capped tubes or bottles containing blood-agar media. Inoculation of blood or pieces of tissue into such media can be used to diagnose infections which are too scanty to be detected by microscopical examination. Strictly sterile procedures must be used at all times, since bacteria and fungi grow readily in such media and trypanosomatids will not usually develop in contaminated tubes.

Trypanosoma and *Leishmania* develop in cultures as the forms normally seen in the arthropod vector and not the vertebrate host (i.e. mainly as promastigote and epimastigote forms, though some species of *Trypanosoma* will develop metacyclic trypomastigote forms). Cultures must be kept at room temperature or 28°C, not hotter. To maintain the parasites in culture, a small drop of fluid from a flourishing culture is transferred aseptically to a tube of fresh medium once or twice a week (a Pasteur pipette, or bacteriological wire loop, is used to transfer the fluid). Details of one blood-agar medium are given below: there are many other equally suitable versions (see Taylor and Baker, 1968).

Dissolve 40 g 'Oxoid' blood agar base no. 2 (Oxo Ltd, Southwark Bridge Road, London SE1, England[1]) in 1 litre of distilled water by steaming. While molten, dispense 5 ml amounts into tubes or bottles and sterilize by autoclaving. When cooled to about 45°C, add aseptically 1 ml of fresh blood to each tube,[2] mix gently and place the tube at an angle of 40–50° so that the agar sets in a slope at the base of the tube. When the agar is set, add (aseptically) 1 ml of the following saline (Locke's solution) to each tube: NaCl, 8 g; KCl, 0·2 g; $CaCl_2$, 0·2 g; KH_2PO_4, 0·3 g; distilled water, 1 litre. (Penicillin, 200 units/ml and streptomycin 200 μg/ml may be added to this solution.)

Incubate the completed tubes at 37°C for about 24 hours to allow diffusion of nutrients into the saline, and also to check for contamination (if not sterile, signs of bacterial growth should be obvious by this time; the presence of any scum or precipitate, either on the agar slope or in the saline, suggests contamination and such tubes should not be used). Tubes may be stored at 4°C for 2–3 weeks if desiccation is prevented by the use of screw caps or rubber caps.

CULTIVATION *in vivo*

Many Protozoa can be grown in tissue-cultures and in avian embryos, but such specialized techniques are seldom resorted to for purposes other than research. The subject has been reviewed by Zuckerman (1966).

Certain species of trypanosomes (including *T. brucei, T. cruzi. T. rangeli* and some strains of *T. congolense* and *T. lewisi*) can be maintained by serial blood passage in laboratory rats or (except *T. lewisi*) in mice. A few drops of blood are collected from the tail of an infected rodent into a syringe containing a saline solution of heparin and, after mixing, the blood suspension is inoculated intraperitoneally to two or three uninfected animals: the frequency with which such passages need to be made depends on the virulence of the strain being used. This procedure needs the provision of adequate facilities and trained personnel, and care must be taken that legislative requirements are met. *Plasmodium berghei* and *Babesia rodhaini* can be maintained similarly in mice, and provide a useful

[1] Other nutrient agars may be used, in the proportions stated by the manufacturers.
[2] Rabbit's blood is most frequently used, but that of horses, sheep, cattle or human beings is usually equally satisfactory. Methods of obtaining blood are described by Taylor and Baker (1968).

source of material for study and demonstration. Species of *Leishmania* and *Toxoplasma gondii* can also be maintained in laboratory rodents but pathogenic organisms such as these should never be handled by persons who have not been specially trained. *L. enriettii*, which does not infect man, can, however, be safely kept in guinea-pigs: the usual procedure is to inoculate the organisms intradermally on the nose of a guinea-pig and, when a lesion has developed, kill the animal and remove the lesion. Impression smears can be made from a cut surface of the lesion, and will contain large numbers of amastigote forms. A portion of the lesion should be chopped or ground up in saline and inoculated to another guinea-pig if it is desired to maintain the strain. This should be done every 3–4 weeks, before the lesion becomes too old and ulcerated; the infection eventually dies out in guinea-pigs.

TECHNIQUES FOR VECTOR HOSTS

The handling and dissection of invertebrate vectors of blood Protozoa will not be dealt with here. Those interested can find details concerning mosquitoes in a book by Shute and Maryon (1966) and concerning tsetse flies in Buxton's monograph (1955).

AVAILABILITY OF MATERIAL FOR STUDY

The trapping and shooting of wild animals is, in many countries, restricted by law and is not, in any case, a thing to be undertaken lightly. Consequently, it is usually best to seek to obtain strains, or preserved material, of the tissue-dwelling Protozoa from other research or teaching laboratories (bearing in mind that such laboratories may not be able to deal with all requests), or from biological suppliers (if such are available). However, subject to the above provisos, blood-films from any freshly killed wild animals are often worthy of examination. Birds may be infected with *Plasmodium, Haemoproteus* or *Leucocytozoon* in most countries (including Britain); trypanosomes may also be present, but they are often too scanty to be detected readily. Wild rabbits, rodents and insectivores may harbour trypanosomes, while the latter two groups may also be hosts to haemogregarines and piroplasms. These parasites, too, are cosmopolitan. Fish, the catching and killing of which are socially sanctioned, may be valuable sources of Cnidospora in all countries of the world. The stickleback *(Gasterosteus aculeatus)* in Britain is commonly infected with a microsporidan, *Glugea anomala*. Fish, too, may be infected with trypanosomes, but parasitaemias are often very low so that parasites may not be easily detectable on thin blood films.

REFERENCES

ASHCROFT, M. T. (1959). *Trop. Dis. Bull.*, **56**, 1073–93
BAKER, J. R. (1961). *Trans. R. Soc. trop. Med. Hyg.*, **55**, 518–24
BAKER, J. R. (1963). *Parasitology*, **53**, 285–92
BAKER, J. R. (1965). In *Evolution of Parasites*, Angela E. R. Taylor (Ed.). Symposium of the British Society for Parasitology, 1–27. Oxford, Blackwell Scientific Publications
BEATTIE, C. P. (1964). *Toxoplasmosis*. Edinburgh, Royal College of Physicians of Edinburgh (publication number 28)
BIRD, R. G. and GARNHAM, P. C. C. (1967). *J. Protozool.*, **14**, Suppl., 42 (Abstract only)
BRAY, R. S. and GARNHAM, P. C. C. (1962). *Indian J. Malariol.*, **16**, 153–5
BROWN, K. N. and BROWN, I. N. (1966). *Trans. R. Soc. trop. Med. Hyg.*, **60**, 2
BÜTTNER, D. W. (1967). *Z. Tropenmed. Parasit.*, **18**, 224–44
BUXTON, P. A. (1955). *The natural history of tsetse flies*. London, H. K. Lewis
CANNING, Elizabeth U., ELKAN, E. and TRIGG, P. I. (1964). *J. Protozool.*, **11**, 157–66
ČERVA, L. and NOVÁK, K. (1968). *Science, N.Y.*, **160**, 92
CHEISSIN, E. M. (1964). *J. Protozool.*, **11**, 91–8
CHEISSIN, E. M. (1965). *Arch. Protistenk.*, **108**, 8–18
CORLISS, J. O. (1959). *J. Protozool.*, **6**, 265–84
CORLISS, J. O. (1960). *Parasitology*, **50**, 111–53

CORLISS, J. O. (1961). *The ciliated Protozoa: characterization, classification and guide to the literature*. Oxford, etc., Pergamon Press

CULBERTSON, C. G., ENSMINGER, P. W. and OVERTON, W. M. (1966). *Amer. J. clin. Path.*, **46**, 305–14

CULBERTSON, C. G., ENSMINGER, P. W. and OVERTON, W. M. (1968). *J. Protozool.*, **15**, 353–63

DANFORTH, W. F. (1967). In *Research in Protozoology*, T. T. Chen (Ed.), **1**, 201–306. Oxford, etc., Pergamon Press

DAVIES, S. F. M., JOYNER, L. P. and KENDALL, S. B. (1963). *Coccidiosis*. Edinburgh and London, Oliver and Boyd

DOBY, J.-M., JEANNES, A. and RAULT, B. (1963). *C. r. hebd. Séanc. Acad. Sci., Paris*, **257**, 248–51

DOGIEL, V. A. (1964). *General Parasitology*, revised and enlarged by Y. I. Polyanski and E. M. Kheisin. (Translated by Z. Kabata.) Edinburgh and London, Oliver and Boyd

DOGIEL, V. A. (1965). *General Protozoology*, 2nd ed., revised and translated by J. I. Poljanskij and E. M. Chejsin. Oxford, Clarendon Press

DRURY, R. A. B. and WALLINGTON, E. A. (1967). *Carlton's histological technique*. New York and Toronto, Oxford University Press

FILICE, F. P. (1952). *Univ. Calif. Publs Zool.*, **57**, 53–146

FRIEDHOFF, K. and SCHOLTYSECK, E. (1968). *Z. ParasitKde.*, **30**, 347–59

FITZPATRICK, J. E. P., KENNEDY, C. C., McGEOWN, Mary C., OREOPOULOS, D. G., ROBERTSON, J. H. and SOYANNWO, M. A. O. (1968). *Nature, Lond.*, **217**, 861–2

GARNHAM, P. C. C. (1966a). *Parasitology*, **56**, 329–34

GARNHAM, P. C. C. (1966b). *Biol. Rev.*, **41**, 561–86

GARNHAM, P. C. C. (1966c). *Malaria parasites*. Oxford, Blackwell Scientific Publications

GARNHAM, P. C. C. (1967). *Protozoology* (Supplement to *J. Helminth.*), **2**, 55–64

GARNHAM, P. C. C. and BRAY, R. S. (1959). *J. Protozool.*, **6**, 352–5

GRASSÉ, P.-P. (Ed.) (1952). *Traité de Zoologie*, **1** (1), Paris, Masson

GRASSÉ, P.-P. (Ed.) (1953). *Traité de Zoologie*, **1** (2), Paris, Masson

GRELL, K. G. (1967). In *Research in Protozoology*, T.-T. Chen (Ed.), **2**, 147–213. Oxford, etc., Pergamon Press

HOARE, C. A. (1956). *Parasitology*, **46**, 130–72

HOARE, C. A. (1959). *Vet. Rev. Annot.*, **5**, 91–102

HOARE, C. A. (1966). *Ergebn. Mikrobiol. ImmunForsch. exp. Ther.*, **39**, 43–57

References

HOARE, C. A. (1967). In *Advances in Parasitology*, B. Dawes (Ed.), **5**, 47–91. London and New York, Academic Press

HOARE, C. A. and WALLACE, F. G. (1966). *Nature, Lond.*, **212**, 1385–6

HONIGBERG, B. M., BALAMUTH, W., BOVEE, E. C., CORLISS, J. O., GOJDICS, M., HALL, R. P., KUDO, R. R., LEVINE, N. D., LOEBLICH, A. R. WEISER, J. and WENRICH, D. H. (1964). *J. Protozool.*, **11**, 7–20

HUNGATE, R. E. (1955). In *Biochemistry and physiology of Protozoa*, S. H. Hutner and A. Lwoff (Eds). 159–99. New York and London, Academic Press

HUTCHISON, W. M., DUNACHIE, J. F. and WORK, K. (1968). *Acta path. microbiol. scandinav.*, **74**, 463–4

HUTNER, S. H. and LWOFF, A. (Eds) (1955). *Biochemistry and Physiology of Protozoa*, **2**. New York and London, Academic Press

JACOBS, L. (1967). In *Advances in Parasitology*, B. Dawes (Ed.), **5**, 1–45. London and New York, Academic Press

JAHN, T. L. and BOVEE, E. C. (1967). In *Research in Protozoology*, T.-T. Chen (Ed.), **1**, 41–200. Oxford etc., Pergamon Press

JAKOWSKA, Sophie and NIGRELLI, R. F. (1956). *Ann. N.Y. Acad. Sci.*, **64** (2), 112–27

KENDALL, S. B. (1959). *Parasitology*, **49**, 169–72

KITCHING, J. A. (1967). In *Research in Protozoology*, T.-T. Chen (Ed.), **1**, 307–36. Oxford etc., Pergamon Press

KUDO, R. R. (1966). *Protozoology*, 5th ed. Springfield, Illinois, Charles C. Thomas

LAINSON, R. (1959). *J. Protozool.*, **6**, 360–71

LAINSON, R., GARNHAM, P. C. C., KILLICK-KENDRICK, R. and BIRD, R. G. (1964). *Brit. med. J.*, ii (5407), 470–2

LAINSON, R. and SHAW, J. J. (1969). *Parasitology*, **59**, 159–62

LEVINE, N. D. (1961). *Protozoan parasites of domestic animals and man*. Minneapolis, Burgess Publishing Company

LOM, J. and CORLISS, J. O. (1967). *J. Protozool.*, **14**, 141–52

LWOFF, A. (Ed.) (1951). *Biochemistry and physiology of Protozoa*, **1**. New York, Academic Press

MACKINNON, Doris L., and HAWES, R. S. J. (1961). *An introduction to the study of Protozoa*. Oxford, Clarendon Press

MAEGRAITH, B. (1948). *Pathological processes in malaria and blackwater fever*. Oxford, Blackwell Scientific Publications

MANWELL, R. D. (1961). *Introduction to Protozoology*. London, Arnold

References

MARTIN, H. M., BARNETT, S. F. and VIDLER, Brenda O. (1964). *Expl Parasit.*, **15**, 527–55

NEAL, R. A. (1966). In *Advances in Parasitology*, B. Dawes (Ed.), **4**, 1–51. London and New York, Academic Press

ORMEROD, W. E. (1963). In *Immunity to Protozoa: Symp. Soc. Brit. Soc. Immunol.*, P. C. C. Garnham, A. E. Previe and I. Roitt (Eds), 213–27. Oxford, Blackwell Scientific Publications

ORMEROD, W. E. (1967). *J. Parasit.*, **53**, 824–30

OXFORD, A. E. (1955). *Expl Parasit.*, **4**, 569–605

PELLÉRDY, L. P. (1965). *Coccidia and coccidiosis*. Budapest, Akademiai Kiado

PITELKA, Dorothy R. (1963). *Electron-microscopic structure of Protozoa*. Oxford etc., Pergamon Press

RICHARDSON, U. F. and KENDALL, S. B. (1963). *Veterinary Protozoology*, 3rd ed. Edinburgh and London, Oliver and Boyd

REID, W. M. (1967). *Expl Parasit.*, **21**, 249–75

RIEK, R. F. (1964). *Aust. J. agric. Res.*, **15**, 802–21

RUDZINSKA, Maria A. and TRAGER, W. (1968). *J. Protozool.*, **15**, 73–88

RUDZINSKA, Maria A. and VICKERMAN, K. (1968). In *Infectious Blood Diseases*, D. Weinman and M. Ristic (Eds), **1**, 217–306. New York and London, Academic Press

SCHUSTER, F. L. (1968). *J. Parasit.*, **54**, 725

SHORTT, H. E. and BLACKIE, E. J. (1965). *J. trop. Med. Hyg.*, **68**, 37–42

SHULMAN, S. S. (1964). *Evolution and phylogeny of Myxosporidia*. Leningrad, Nauka

SHUTE, P. G. and MARYON, Marjorie E. (1966). *Laboratory technique for the study of malaria*, 2nd ed. London, J. and A. Churchill

SMITH, B. F. and STEWART, Babette T. (1966). *Expl Parasit.*, **19**, 52–63

SPRAGUE, V. and VERNICK, S. H. (1968). *J. Protozool.*, **15**, 547–71

SPRENT, J. F. A. (1963). *Parasitism*. London, Baillière, Tindall and Cox

STEBHENS, W. E. and JOHNSTON, M. R. L. (1966). *J. Ultrastruct. Res.*, **15**, 543–54

TAYLOR, Angela E. R. and BAKER, J. R. (1968). *The cultivation of parasites in vitro*. Oxford, Blackwell Scientific Publications

TRUSSELL, R. E. (1947). *Trichomonas vaginalis and trichomoniasis*. Springfield, Illinois, Charles C. Thomas and Oxford, Blackwell Scientific Publications

VICKERMAN, K. (1965). *Nature, Lond.*, **208**, 762–6

VOLLER, A. (1964). *Bull. Wld Hlth Org.*, **30**, 343–54

References

WALLIKER, D. (1968). *J. Protozool.*, **15**, 571–5

WANG, S. S. and FELDMAN, H. A. (1967). *New Engl. J. Med.*, **277**, 1174–9

WEINMAN, D. and RISTIC, M. (Eds) (1968). *Infectious blood diseases of man and animals: diseases caused by Protista*, **1**. New York and London, Academic Press

WENYON, C. M. (1926). *Protozoology*. London, Baillière, Tindall and Cox. Reprinted 1966 by Baillière, Tindall and Cassel

WESSENBERG, H. (1961). *Univ. Calif. Publs Zool.*, **61**, 315–70

WHITE, W. F., SAXTON, H. M. and DAWSON, I. M. P. (1961). *Brit. med. J.*, ii (5263), 1327–31

WORK, K. and HUTCHISON, W. M. *Acta path. microbiol. scandinav.*, **75**, 191–192

ZUCKERMAN, Avivah (1966). *Ann. N.Y. Acad. Sci.*, **139**, 24–38

INDEX

Page numbers in italics refer to illustrations

ABLASTIN, 51
Acanthamoeba, 85
Adelea, 19, 26, *89*
Adeleina, 19, 24, 25, 90, 91, 96
Aegyptianella, 126
Aggregata, 93
Akiba caulleryi, 116
Allantosoma, 21, 150
Amastigotes, 42, 43, 44, 52, 164
Amoebida, 18, 77
Amphibia, 39, 45, 65, 69, 71, 75, 80, 92, 117, 125, 142, 146
Anaplasma, 126
Angeiocystis, 93
Anisogamete, 39
Annelids, 23, 89, 90, 93, 137, 145
Anopheles, 101, 105, 113
Apostomatida, 145
Archigregarinida, 89, 90
Axoneme, 30, 102
Axostyle, 31, 66

Babesia, 19, 118, 120, 122
 B. argentina, 119
 B. bigemina, 119, 121
 B. bovis, 119, 120
 B. caballi, 119
 B. canis, *118*, 119
 B. divergens, *118*, 119, 120
 B. equi, 119
 B. felis, 119
 B. gibsoni, 119
 B. major, 119
 B. motasi, 119
 B. ovis, 119
 B. rodhaini, 120, 163
 B. trautmanni, 119
Babesiasis, 120–2
Babesiidae, 19, 118
Babesiosoma, 125
Balantidiosis, 147, 148
Balantidium, 20, 146–9, 157
 B. coli, 146, 147, *147*, 156
Barrouxia, 93
Bartonella, 126
Besnoitia, 19, 133
 B. bennetti, 133
 B. besnoiti, *129*, 133
 B. jellisoni, 133
Besnoitiidae, 127, 133
Birds, 45, 65, 66, 69, 71, 73, 92, 100, 101, 102, 114, 115, 116, 117, 118, 128, 157, 164
Blackhead, 65
Blackwater fever, 112
Blastocrithidia, 42
Blepharocorys, 150
Blood films, 158–61, *159*
Bodonina, 25, 40
Boophilus, 119, 123
Buetschlia, 20, 150
Bufo bufo, 143

CAMELS, 54, 55, 146, 149
Carinamoeba, 101
Carnoy's fluid, 162
Caryospora, 93
Caryotropha, 93
Cats, 45, 50, 51, 69, 80, 95, 119, 132

Cattle, 71, 73, 80, 119, 120, 123, 124, 130, 132, 133, 147, 149
Ceratomyxa, 139
Ceratophyllus fasciatus, 50
Ceratopogonidae, 115
Chagas's disease, 51–2
Charonina, 150
Chickens, 73, 80, 114, 116, 133, 157
Chilomastix, 17
　C. mesnili, 67, 67, 68
Chimpanzee, 51, 108, 113
Chonotrichida, 145
Chromatoid bodies, 78, 80, 81, 153
Cilia, 29–30, 33–4, 146, 147, 148
Ciliophora, 22, 25, 28, 29, 31, 32, 34, 37, 38, 39
Classification, of Phylum Protozoa, 17–21
Cnidospora, 19, 22, 25, 28, 31, 98, 137, 164
Coccidia, 24, 25, 31, 34, 38, 87, 88, 90, 92, 154
Commensalism, 12, 149
Complement-fixation reaction, 132, 135
Contractile vacuoles, 32, 148
Crithidia, 41
Cryptobia, 17, 40
　C. keysselitzi, 41
Cryptosporidium, 93
Ctenodactylus gundi, 128
Culicidae, 100, 101
Culicoides, 115, 116
Cultures, 35, 45, 63, 65, 72, 149, 162–4
Cutaneous leishmaniasis, 44
Cyclospora, 93
Cysts, 31, 39, 77, 80, 82, 130, 134, 135, 142, 148, 153, 154
Cytauxzoon, 19, 122
Cytochrome system, 36
Cytomeres, 38, 122
Cytopharynx, 35
Cytopyge, 37, 148
Cytostome, 35, 38, 66, 148

Dactylosoma, 19, 125
Dactylosomidae, 125
Dasytricha, 20, 150
Dermacentor, 119
Dientamoeba, 17, 65, 76
　D. fragilis, 77, 79, 84
Diplodinium, 21, 150, 151
Diplomonadida, 69
Diptera, 50, 100, 101, 115, 116
Dogs, 44, 45, 50, 51, 54, 55, 69, 80, 95, 119, 120, 130, 132, 135, 147

Donkeys, 54, 55, 62
Dorisiella, 93
Ducks, 115, 116, 133
Duttonella, 47, 56
Dye test, 132

Echinospora, 93
Echinozoon, 118
Eimeria, 19, 26, 31, 92–7, 157
　E. adenoeides, 95
　E. ahsata, 95
　E. arloingi, 95
　E. bovis, 95
　E. bucephalae, 95
　E. canis, 95
　E. debliecki, 95
　E. irresidua, 95
　E. magna, 95
　E. meleagrimitis, 95
　E. necatrix, 95
　E. perforans, 96
　E. stiedae, 95, 96
　E. tenella, 95
　E. truncata, 95
　E. zurnii, 95
Eimeriidae, 91, 92, 97, 102
Eimeriina, 19, 24, 25, 90, 92, 96, 97, 98, 101
Eimeriinae, 93
Embden-Meyerhof pathway, 36
Encephalitozoon, see *Nosema*
Endamoeba, 18, 76
Endolimax, 18, 76
　E. nana, 79, 84
Endotrypanum, 41, 42
Entamoeba, 18, 35, 76–80, 153
　E. bovis, 80
　E. chattoni, 80
　E. coli, 79, 84
　E. gallinarum, 80
　E. gingivalis, 77
　E. hartmanni, 79, 80, 84, 85
　E. histolytica, 12, 22, 71, 76, 79, 80–5, 148, 156, 157
　E. invadens, 76, 80
　E. moshkovskii, 80
　E. muris, 80
　E. ovis, 80
　E. polecki; see *E. suis*
　E. ranarum, 80
　E. suis, 80, 157
Enteromonas hominis, 68
Entodinium, 21, 150
Eperythrozoon, 126
Epimastigote, 42, 49, 52, 162
Equidae, 119, 149, 150
Espundia, 44

Eucoccida, 38, 90, 102
Eugregarinida, 38, 88, 90
Evolution, of Protozoa, 21–6
Excretion, 37
Exflagellation, 102
Exoerythrocytic stages, Malaria, 101–5, 108, 110, 111, 115; Theileriidae, 122, 124

FAECAL EXAMINATIONS, 152–5
Fibrils, 29, 146
Fish, 45, 69, 75, 94, 98, 117, 125, 138, 140, 141, 142, 145, 146, 164
Flagellum, 29–30, 41, 42, 43, 66
Fluorescent antibodies, 106, 112
Frogs, 147, 157

GAMETE, 88, 91, 94
Gametocyst, 88
Gametocyte, 24, 88, 91, 94, 100, 102, 114, 115, 125
Gametogony, 91, 94, 102
Gasterosteus aculeatus, 164
Giardia, 17, 28, 64, 69, 70, 157
 G. canis, 69
 G. intestinalis, 69
 G. lamblia, 67, 68, 69, 70
 G. muris, 69
Giemsa's stain, 45, 63, 72, 106, 156, 160, 161, 162
Giovannolaia, 101
Glossina morsitans, 61
 G. palpalis, 61
Glossinidae, 52, 54, 55, 57, 58, 59, 60, 61
Glugea, 20
 G. anomala, 164
Goats, 119, 123, 124, 149
Gregarina ovata, 89
Gregarinia, 24, 25, 28, 34, 88
Gymnostomatida, 145, 150

Haemamoeba, 101, 114
Haemaphysalis, 119
Haematoxylin stain, 155, 162
Haemobartonella, 126
Haemogregarina, 19, 34, 89, 90, 91, 92, 164
Haemogregarinidae, 90
Haemoproteidae, 100, 114, 115
Haemoproteus, 19, 115, 164
 H. palumbis, 114, 115
Haemosporina, 19, 24, 25, 26, 34, 38, 90, 96, 97, 100, 104, 117
Haemozoin (malaria pigment), 37, 38
Haplosporea, 25, 87, 98, 99

Hartmanella, 18, 22, 32, 79, 85, 86
Hartmanellidae, 76
Hemiptera, 42, 50, 63
Henneguya salminicola, 138, 139
Hepatocystis, 19, 115
Hepatozoidae, 90, 91
Hepatozoon, 91
 H. balfouri, 89
 H. muris, 91
Herpetomonas, 41
Herpetosoma, 47, 50
Heterakis gallinae, 65, 66
Hexamita, 17, 64, 69, 157
 H. columbae, 69
 H. meleagridis, 69
 H. salmonis, 69
Hippoboscidae, 50, 115
Histomonas, 64, 65–6, 77
 H. meleagridis, 17, 65, 66, 67, 156
Hoarella, 93
Holotrichia, 145, 146, 150
Horses, 54, 55, 62, 80, 133
Hosts, 13
Huffia, 101
Hyalomma, 119, 123
Hypermastigida, 12, 23, 39, 64, 74

Ichthyophthirius, 20, 145
Iodamoeba, 18, 76
 I. buetschlii, 79, 84, 157
Isoptera (Termites), 12, 23, 28, 64, 69, 71, 74, 149
Isospora, 19, 31, 93, 98, 157
 I. belli, 92, 98
 I. bigemina, 95, 98
 I. felis, 95
 I. hominis, 92, 98, 98
 I. suis, 95
Isotricha, 20, 150
Ixodes, 119

KALA-AZAR, 44
Karyolysus, 19, 90, 91
Kinetoplast, 40, 41, 47, 48, 50, 56
Kinetoplastida, 40, 45, 64
Kinetosome, 30
Klossia, 19
Klossiella, 19
Koch's blue bodies, 124
Krebs's cycle, 36, 60
Kudoa, 20, 140

Lamblia intestinalis, see *Giardia lamblia*
Lankesterella, 19, 92, 93
 L. garnhami, 89

Laverania, 101
Leeches, 24, 41, 45, 90, 91, 92
Leishmania, 17, 23, 24, 25, 42, 43–5, 162, 164
  L. braziliensis, *41*, 44
  L. donovani, 44, 45
  L. enriettii, 45, 164
  L. tropica, 28, *41*, 44, 45
    L. t. mexicana, *41*
Leishmaniasis, 44–5
Leptomonas, 17, 41
Leucocytozoidae, 100, 115
Leucocytozoon, 19, 116, 161, 164
  L. simondi, *114*, 116
Limiting membrane, 31
Lizards, 45, 116, 133
Locke's solution, 163
Locomotion, 32–4
Lophomonadina, 75

MALARIA, 87, 105, 108, 110–13, 120, 121
Malaria parasites, 13, 16, 26, 29, 31, 32, 37, 100, 101, 117
Mammals, 24, 36, 43, 46, 49, 51, 58, 60, 69, 76, 92, 100, 101, 115, 117, 118, 128, 142, 144, 158
Man, 44, 45, 50, 51, 55, 61, 67, 69, 71, 76, 80, 82, 83, 85, 86, 92, 94, 98, 105, 108, 110–13, 117, 132, 133, 134, 142, 144, 147, 148, 149, 158
Mantonella, 93
Mastigophora, 15, 28, 29, 30, 31, 32, 34, 37, 64, 75
Maurer's clefts, 107, 110
Median body, 70
Megatrypanum, 47, 49
Merocyst, 115
Merocystis, 93
Merozoite, 38, 88, 94, 101, 102, 104, 124
Metatrypanosomes, 45, 49
Microsporida, 20, 38, 142, 143
Microsporidea, 20, 142
Microtubules, 29, 72
Midges, 115
Mites, 90, 92
Mitochondria, 28–9, 36, 38, 40, 60
Molluscs, 23, 71, 92, 137, 145
Monkeys, 50, 51, 54, 57, 69, 71, 85, 113, 115, 133, 147, 156
Monocystis, 18, 90
Mosquito, 13, 100, 101, 102, 105, 142, 164
Mucocutaneous leishmaniasis, 44
Myriospora, 93

Myxobolus, 20, 138, 139, 140, *141*
Myxosoma cerebralis, 138, 141
  M. heterospora, 140
Myxosporida, 137–41
Myxosporidea, 19, 137

Naegleria, 22, 32, *79*, 85
Nannomonas, 47, 56
Nematodes, 23, 65, 77
Neogregarinida, 18, 89, 90
Nosema, 20
  N. apis, 142, 143
  N. bombycis, 142, 143
  N. cuniculi, *141*, 142, 144
Novyella, 101
Nucleus, 28, 38, 47, 48, 50
Nutrition, 34–6
Nuttallia, 120
Nyctotherus, 21, 157
  N. cordiformis, 147
  N. ovalis, 147

Ochoterenaia, 150
Octosporella, 93
Oocyst, 88, 89, 91, 92, 93, 94, 96, 104, 154
Ookinete, 91, 102, 121
Opalina, 18, 75, 147, 157
Opalinata, 25, 28, 34, 39, 64, 75
Ophidiella, 101
Ophryoscolecidae, 149
Ophryoscolex caudatus, 21, *147*, 150
Opossums, 50, 51, 146
Oriental sore, 41, 44
Orthoptera, 28, 64, 69, 74
Ovivora, 93
Oxymonadida, 17, 64, 69

PANSPOROBLAST, 139
Parabasal body, 72
Parahaemoproteus, 115
Paraisotricha, 150
Paramoeba, 76
Parasitism, 11, 12, 145
Pellicle, 31, 34, 35, 36
Peritrichia, 21, 146
Peritrophic membrane, 60
Permanent preparations, 155
Pfeiferinella, 93
Phagotrophy, 34, 35
Phlebotomus, 43
Phoresey, 11
Phytomastigophorea, 17, 25
Phytomonas, 17, 24, *41*, 42
Pigeons, 73–4, 115

Pigs, 54, 55, 57, 69, 80, 119, 132, 133, 146–9, 157
Pirhemocyton, 126
Piroplasms, 16, 117–26, 164
Plasmalemma, 31, 107
Plasmodiidae, 100, 115
Plasmodium, 19, 25, 26, 28, 29, 31, 34, 35, 38, 100–15, 121, 164
  P. berghei, 29, 114, 163
  P. cynomolgi, *103*, 113
  P. falciparum, 26, 101, 104, 106, 108–12, *109*, 121
  P. gallinaceum, 114, *114*
  P. gonderi, 113
  P. knowlesi, 113
  P. malariae, 106, 107, 108, *109*, 110, 113
  P. ovale, 106, 108, *109*, 110, 113
  P. vivax, *103*, 105, 106, 107, 113
Pleomorphism, 58, 61
Plistophora, 20
  P. myotrophica, 143
Pneumocystis carinii, *129*, 135–6
Pneumonia, atypical interstitial plasma-cell, 135–6
Post kala-azar dermal leishmanoid, 44
Primates, 80, 101, 113, 115, 146
Promastigotes, 42, 162
Protococcida, 90
Pseudocyst, 127, 130, 131, 133, 134, 142, 144
Pseudoklossia, 93
Psychodidae, 43
Pycnomonas, 47, 57
Pythonella, 93

Rabbits, 51, 54, 55, 96, 133, 144, 157, 164
Reduviidae, 50, 63
Reproduction, 37–9; Binary fission, 37–8, 43, 65, 124, 133, 135, 148; Budding, 38; Endodyogeny, 38–9; Multiple fission, 38, 124; Schizogony, 26, 38, 88, 90, 91, 92, 94, 100, 101, 122; Conjugation, 39, 146, 148
Reptiles, 45, 69, 71, 75, 76, 80, 92, 100, 101, 102, 114, 115, 118, 125
Respiration, 36–7
Retortamonadida, 17, 66–7
Retortamonas hominis, 67
  R. intestinalis, 67, 68
Rhipicephalus, 119, 123
Rhizomastigida, 17, 64
Rhizopodea, 18, 33, 35, 36, 37, 76
Rhodnius prolixus, 50, 51
Rhynchoidomonas, 42

Rodents, 44, 45, 50, 51, 54, 55, 58, 69, 80, 114, 118, 120, 128, 133, 135, 144, 157, 164
Ruminants, 49, 50, 54, 55, 56, 71, 122, 149, 150, 151

Salivaria, 24, 25, 46, 49, 50, 52, 63, 162
Salmonella, 148
Sarcocystidae, 127, 133
Sarcocystis, 19, 31, 128, 133, 134
  S. lindemanni, *129*, 134, 135
  S. muris, 135
  S. tenella, 134
Sarcodina, 22, 25, 27, 32, 34, 76, 117
Sarcomastigophora, 17, 21, 25, 64
Sargeaunt's stain, 154
Sauramoeba, 101
Saurocytozoon, 116
Schaudinn's solution, 155
Schellackia, 92, 93
Schizont, 38, 94, 101, 110, 125
Schizotrypanum, 47, 51
Schüffner's dots, 106, 107, 108, 110
Secretory organelles, 29
Sheep, 54, 55, 56, 80, 119, 123, 124, 130, 132, 133, 134, 146, 147, 149
Simulium, 116
Skeletal structures, 31
Sleeping sickness, 61, 62
Sphaerospora tincae, 141
Spirotrichia, 145, 150, 151
Sporadin, 88
Spore, 137–44
Sporoblast, 94, 139
Sporocyst, 89, 91, 93
Sporogony, 26, 91, 92, 96, 99, 104, 108
Sporont, 139, 142
Sporoplasms, 137, 139, 142
Sporozoa, 16, 18, 22, 24, 25, 26, 28, 32, 34, 35, 37, 38, 39, 87, 98, 100, 117, 127
Sporozoite, 24, 87, 92, 93, 94, 101, 104
Stercoraria, 46, 49, 52
Suctoria, 146, 150
Symbiosis, 12–13, 74
Syzygy, 88, 90, 91, 92

Tabanidae, 50
Tabanus, 56
Telomyxa glugeiformis, 143
Telosporea, 18, 87
Tetrahymena, 20, 22, 145
Theileria, 19, 122–4

T. annulata, 123–5
T. hirci, 123, 124
T. lawrencei, 124, 125
T. mutans, 123, 125
T. ovis, 123
T. parva, *118*, 123, 124, 125
Theileriasis, 124–5
Theileriidae, 122
Thelohania apodemi, 20, 144
T. legeri, 142
Thyrsites atum, 140
Ticks, 90, 117, 120, 121, 122, 124
Toxoplasma, 19, 31, 35, 65, 127, *129*, 134
T. gondii, 128, *129*, 133, 164
T. hominis, 127
T. microti, 128
Toxoplasmatidae, 127, 133
Toxoplasmea, 19, 25, 34, 38, 87, 127
Toxoplasmosis, 128, 130–3
Transmission, 13
Trichodina, 21, 146
Trichomonadida, 71
Trichomonas, 18, 31, 64, 65, 71–4, 157
T. foetus, 71, 73
T. gallinae, 73–4
T. hominis, 68, 71
T. muris, 67
T. tenax, 71
T. vaginalis, 71, 72, 73
Trichonympha, 18, 74
T. campanula, 67
Trichonymphina, 75
Trichostomatida, 20, 145, 146, 150
Trypanosoma, 17, 23, 42, 45–63, 161, 162, 164
T. avium, 46
T. brucei sspp., 36, 47, 48, 55, 58, *59*, 60, 61, 62, 111, 163
T. b. brucei, 55, 58, 60, 61
T. b. gambiense, 55, 58, 60, 61, 63
T. b. rhodesiense, 12, 55, 58, 60, 61, 63
T. congolense, 47, 48, 54, *57*, 57, 163
T. cruzi, 23, 25, 28, 47, 48, 51, 52, *53*, 63, 163
T. dimorphon, 57
T. equinum, 47, 55, 58, *59*, 61, 62
T. equiperdum, 47, 48, 55, 58, 61, 62
T. evansi evansi, 28, 47, 48, 55, 58, *59*, 61, 62
T. lewisi, 12, 28, *46*, 47, 50, 51, 52, 53, 163
T. melophagium, 47
T. microti, 51
T. musculi, 50, 51
T. nabiasi, 51
T. primatum, 51
T. rangeli, 42, *46*, 47, 48, 49, 50, 163
T. rotatorium, 45, *46*
T. simiae, 47, 48, 54, *57*, 57
T. suis, 47, 48, 54, *57*, 57
T. theileri, *46*, 47, 48, 50
T. uniforme, 47, 48, 54, 56, *57*
T. vivax viennei, 47, 54, 56
T. v. vivax, 47, 48, 54, 56, *57*, 61
Trypanosomatina, 23, 25, 38, 40, 41
Trypanosomiasis, see Sleeping sickness
Trypanozoon, 28, 47, 58
Trypomastigote, 42, 49, 50, 52, 56, 162
Tsetse flies, 53, 56, 57, 58, 60, 61, 62, 164
Turkeys, 73
Tyzzeria, 93, 95

Vinckeia, 101
Visceral leishmaniasis, 44

Wenyonella, 93

XENODIAGNOSIS, 63
Xenopsylla cheopis, 50

Yakimovella, 93

ZIEMANN'S DOTS, 107, 108, 110
Zoite, 127, 129, 130
Zoomastigophorea, 17, 35, 38, 40, 64, 68